T0310151

FIBER OPTIC ESSENTIALS

The Wiley Bicentennial–Knowledge for Generations

Each generation has its unique needs and aspirations. When Charles Wiley first opened his small printing shop in lower Manhattan in 1807, it was a generation of boundless potential searching for an identity. And we were there, helping to define a new American literary tradition. Over half a century later, in the midst of the Second Industrial Revolution, it was a generation focused on building the future. Once again, we were there, supplying the critical scientific, technical, and engineering knowledge that helped frame the world. Throughout the 20th Century, and into the new millennium, nations began to reach out beyond their own borders and a new international community was born. Wiley was there, expanding its operations around the world to enable a global exchange of ideas, opinions, and know-how.

For 200 years, Wiley has been an integral part of each generation's journey, enabling the flow of information and understanding necessary to meet their needs and fulfill their aspirations. Today, bold new technologies are changing the way we live and learn. Wiley will be there, providing you the must-have knowledge you need to imagine new worlds, new possibilities, and new opportunities.

Generations come and go, but you can always count on Wiley to provide you the knowledge you need, when and where you need it!

William J. Pesce
President and Chief Executive Officer

Peter Booth Wiley
Chairman of the Board

FIBER OPTIC ESSENTIALS

K. Thyagarajan
Ajoy Ghatak

IEEE PRESS

WILEY-
INTERSCIENCE

A JOHN WILEY & SONS, INC., PUBLICATION

Published by John Wiley & Sons, Inc., Hoboken, New Jersey.
Published simultaneously in Canada.

Wiley Bicentennial Logo: Richard J. Pacifico

Library of Congress Cataloging-in-Publication Data:

Thyagarajan, K.
Fiber optic essentials / by K. Thyagarajan, Ajoy Ghatak.
p. cm.
Includes bibliographical references.
ISBN 978-0-470-09742-7
1. Fiber optics. 2. Optical fiber communication equipment and supplies.
I. Ghatak, A. K. (Ajoy K.), 1939– II. Title.
TA1800.T49 2007
621.36′92–dc22 2006103090

Printed in the United States of America

10 9 8 7 6 5 4 3 2 1

To Raji and Gopa

CONTENTS

PREFACE

The dramatic reduction in transmission loss of optical fibers coupled with very important developments in the area of light sources and detectors have resulted in phenomenal growth of the fiber optic industry during the last 35 years or so. Indeed, the birth of optical fiber communication systems coincided with the fabrication of low-loss optical fibers and the operation of room-temperature semiconductor lasers in 1970. Since then, scientific and technological growth in this field has been phenomenal. Although the major applications of optical fibers have been in the area of telecommunications, many new areas, such as fiber optic sensors, fiber optic devices and components, and integrated optics, have witnessed immense growth.

As with any technological development, the field of fiber optics has progressed through a number of ideas based on sound mathematical and physical principles. For a thorough understanding of these, one needs to go through a good amount of mathematical rigor and analysis, which is carried out in undergraduate and graduate curricula. At the same time there are a sizable number of engineering and technical professionals, technical managers, and inquisitive students of other disciplines who are interested in having a basic understanding of various aspects of fiber optics either to satisfy their curiosity or to help them in their professions. For these professionals a book describing the most important aspects of fiber optics without too much mathematics, based purely on physical reasoning and explanations, should be very welcome. A book taking the reader from the basics to the current state of development in fiber optics does not seem to exist, and the present book aims to fill that gap.

The book begins with a basic discussion of light waves and the phenomena of refraction and reflection. The next set of chapters introduces the reader to the field of fiber optics, discussing different types of fibers used in communication systems, including dispersion-compensating fibers. In later chapters we discuss recent developments, such as fiber Bragg gratings, fiber amplifiers, fiber lasers, nonlinear fiber optics, and fiber optic sensors. Examples and comparison with everyday experience are provided wherever feasible to help readers understanding by relation to known facts. The book is interspersed with numerous diagrams for ease of visualization of some of the concepts.

The mathematical details are kept to a bare minimum in the hope of providing easy reading and understanding of some of the most important technological developments of the twentieth century, which are penetrating more and more deeply into our society and helping to make our lives a bit easier.

We are very grateful to all our colleagues and students at IIT Delhi for numerous stimulating discussions and academic collaborations. One of the authors (A.G.) is grateful to Disha Academy of Research and Education, Raipur for supporting this endeavor.

K. THYAGARAJAN
AJOY GHATAK

New Delhi

UNITS AND ABBREVIATIONS

1 Å (1 angstrom)	one-tenth of a billionth of a meter (= 10^{-10} m)
1 nm (1 nanometer)	one-billionth of a meter (= 10^{-9} m)
1 μm (1 micrometer)	one-millionth of a meter (= 10^{-6} m)
1 cm (1 centimeter)	one-hundredth of a meter (= 10^{-2} m)
1 mm (1 millimeter)	one-thousandth of a meter (= 10^{-3} m)
1 km (1 kilometer)	1000 meters (= 10^{3} m)
speed of light in vacuum, *c*	300 million kilometers per second (= 3×10^{8} m/s)
1 fs (1 femtosecond)	one-millionth of a billionth of a second (= 10^{-15} s)
1 ps (1 picosecond)	one-thousandth of a billionth of a second (= 10^{-12} s)
1 ns (1 nanosecond)	one-billionth of a second (= 10^{-9} s)
1 μs (1 microsecond)	one-millionth of a second (= 10^{-6} s)
1 ms (1 millisecond)	one-thousandth of a second (= 10^{-3} s)
1 kHz (1 kilohsertz)	1000 vibrations per second (= 10^{3} Hz)
1 MHz (1 megahertz)	1 million vibrations per second (= 10^{6} Hz)
1 GHz (1 gigahertz)	1 billion vibrations per second (= 10^{9} Hz)
1 THz (1 terahertz)	1000 billion vibrations per second (= 10^{12} Hz)
1 nW (1 nanowatt)	one-billionth of a watt (= 10^{-9} W)
1 μW (1 microwatt)	one-millionth of a watt (= 10^{-6} W)
1 mW (1 milliwatt)	one-thousandth of a watt (= 10^{-3} W)
1 kW (1 kilowatt)	1000 watts (= 10^{3} W)
1 MW (1 megawatt)	1 million watts (= 10^{6} W)
3 dB loss	power loss by a factor of 2
10 dB loss	power loss by a factor of 10
20 dB loss	power loss by a factor of 100
30 dB loss	power loss by a factor of 1000
3 dB gain	power amplification by a factor of 2
10 dB gain	power amplification by a factor of 10
20 dB gain	power amplification by a factor of 100
30 dB gain	power amplification by a factor of 1000
1 kb/s	1000 bits per second (= 10^{3} bits per second)
1 Mb/s	1 million bits per second (= 10^{6} bits per second)
1 Gb/s	1 billion bits per second (= 10^{9} bits per second)
1 Tb/s	1000 billion bits per second (= 10^{12} bits per second)
0 dBm	1 mW
−30 dBm	1 μW
+30 dBm	1 W

AM	amplitude modulation
APD	avalanche photo diode
ASE	amplified spontaneous emission
AWG	arrayed waveguide grating
BER	bit error rate
BW	bandwidth
CSF	conventional single-mode fiber
CW	continuous wave
CWDM	coarse wavelength-division multiplexing
dB	decibel
DBR	distributed Bragg reflector
DCF	dispersion-compensating fiber
DFB	distributed-feedback
DMD	differential mode delay
DSF	dispersion-shifted fiber
DWDM	dense wavelength-division multiplexing
EDFA	erbium-doped fiber amplifier
FBG	fiber Bragg grating
FM	frequency modulation
FOG	fiber optic gyroscope
FSO	free-space optics
FTTH	fiber to the home
FWM	four-wave mixing
ITU	International Telecommunication Union
LD	laser diode
LEAF	large effective area fiber
LED	light-emitting diode
LPG	long-period grating
MCVD	modified chemical vapor deposition
MZ	Mach–Zehnder
NA	numerical aperture
NEP	noise equivalent power
NF	noise figure
NRZ	non return to zero
NZDSF	nonzero dispersion-shifted fiber
OOK	on–off keying
OSNR	optical signal-to-noise ratio
OTDR	optical time-domain reflectometer
PCM	pulse-code modulation
PIN	p(doped)–intrinsic–n(doped)
PMD	polarization mode dispersion
RFA	Raman fiber amplifier
RZ	return to zero
SC	supercontinuum

SDH	synchronous digital hierarchy
SMF	single-mode fiber
SNR	signal-to-noise ratio
SOA	semiconductor optical amplifier
SONET	synchronous optical network
SPM	self-phase modulation
TDM	time-division multiplexing
TIR	total internal reflection
VCSEL	vertical cavity surface-emitting laser
XPM	cross-phase modulation
WDM	wavelength-division multiplexing

CHAPTER 1

Introduction

Optics today is responsible for many revolutions in science and technology. This has been brought about primarily by the invention of the laser in 1960 and subsequent development in realizing the extremely wide variety of lasers. One of the most interesting applications of lasers with a direct impact on our lives has been in communications. Use of electromagnetic waves in communication is quite old, and development of the laser gave communication engineers a source of electromagnetic waves of extremely high frequency compared to microwaves and millimeter waves. The development of low-loss optical fibers led to an explosion in the application of lasers in communication, and today we are able to communicate almost instantaneously between any two points on the globe. The backbone network providing this capability is based on optical fibers crisscrossing the Earth: under the seas, over land, and across mountains. Today, more than 10 terabits of information can be transmitted per second through one hair-thin optical fiber. This amount of information is equivalent to simultaneous transmission of about 150 million telephone calls—certainly one of the most important technological achievements of the twentieth century. We may also mention that in 1961, within one year of the demonstration of the first laser by Theodore Maiman, Elias Snitzer fabricated the first fiber laser, which is now finding extremely important applications in many diverse areas: from defense to sensor physics.

Since fiber optic communication systems are playing very important roles in our lives, an introduction to these topics, with a minimum amount of mathematics, should give many interested readers a glimpse of the developments that have taken place and that continue to take place. In Chapter 2 we introduce the reader to light waves and their characteristics and in Chapter 3 explain how it is possible to use light waves to carry information. Chapters 4 to 8 deal with various characteristics of the optical fiber relevant for applications in communication and sensing. The erbium-doped fiber amplifier has revolutionized high-speed communication; this is discussed in Chapter 9, where we also discuss fiber lasers, which have found extremely important industrial applications. Chapter 10 covers Raman fiber amplifiers, which are playing increasingly important roles in optical communication systems. In Chapter 11 we describe fiber Bragg grating, which is indeed a very beautiful device with numerous practical

Fiber Optic Essentials, By K. Thyagarajan and Ajoy Ghatak

applications. In Chapter 12 we discuss some important fiber optic components, which are an integral part of many devices used in fiber optic communication systems.

When the light power within an optical fiber becomes substantial, the properties of the fiber change due to the high intensity of the light beam. Such an effect, called a nonlinear effect and discussed in Chapter 13, plays a very important role in the area of communication. There is also considerable research and development (R & D) effort to utilize such effects for signal processing of optical signals without converting them into electronic signals. Such an application should be very interesting when the speed of communications that use light waves goes up even further as electronic circuits become limited due to the extremely fast response required. Fiber optic sensors, discussed in Chapter 14, form another very important application of optical fibers, and some of the sensors discussed are already finding commercial applications. They are expected to outperform many conventional sensors in niche applications and there is a great deal of research effort in this direction.

In this book we introduce and explain various concepts and effects based on physical principles and examples while keeping the mathematical details to a minimum. The book should serve as an introduction to the field of fiber optics, one of the most important technological revolutions of the twentieth century. If it can stimulate the reader to further reading in this exciting field and help him or her follow developments as they are taking place, with applications in newer areas, it will have served its purpose.

CHAPTER 2

Light Waves

2.1 INTRODUCTION

What is light? That is indeed a very difficult question to answer. To quote Richard Feynman: "Newton thought that light was made up of particles, but then it was discovered that it behaves like a wave. Later, however (in the beginning of the twentieth century), it was found that light did indeed sometimes behave like a particle. . . . So it really behaves like neither." However, all phenomena discussed in this book can be explained very satisfactorily by assuming the wave nature of light. Now the obvious question is: What is a wave? A *wave* is propagation of disturbance. When we drop a small stone in a calm pool of water, a circular pattern spreads out from the point of impact (Fig. 2.1).[1] The impact of the stone creates a disturbance that propagates outward. In this propagation, the water molecules do not move outward with the wave; instead, they move in nearly circular orbits about an equilibrium position. Once the disturbance has passed a certain region, every drop of water is left at its original position. This fact can easily be verified by placing a small piece of wood on the surface of water. As the wave passes, the piece of wood comes back to its original position. Further, with time, the circular ripples spread out; that is, the disturbance (which is confined to particular region at a given time) produces a similar disturbance at a neighboring point slightly later, with the pattern of disturbance remaining roughly the same. Such a propagation of disturbances (without any translation of the medium in the direction of propagation) is termed a *wave*. Also, the wave carries energy; in this case the energy is in the form of the kinetic energy of water molecules. There are many different types of waves: sound waves, light waves, radio waves, and so on, and all waves are characterized by properties such as wavelength and frequency.

2.2 WAVELENGTH AND FREQUENCY

We next consider the propagation of a transverse wave on a string. Imagine that you are holding one end of a string, with the other end being held tightly by another

[1] Water waves emanating from a point source are shown very nicely at the Web site http://www.colorado.edu/physics/2000/waves_particles/waves.html.

Fiber Optic Essentials, By K. Thyagarajan and Ajoy Ghatak

FIGURE 2.1 Water waves spreading out from a point source. (Adapted from http://www.colorado.edu/physics/2000/waves_particles/waves.html.)

person so that the string does not sag. If we move the end of the string in a periodic up-and-down motion ν times per second, we generate a wave propagating in the $+x$ direction. Such a wave can be described by the equation (Fig. 2.2)

$$y(x, t) = a \sin(\omega t - kx) \tag{2.1}$$

where a and ω ($=2\pi\nu$) represent the amplitude and angular frequency of the wave, respectively; further,

$$\lambda = \frac{2\pi}{k} \tag{2.2}$$

represents the wavelength associated with the wave. Since the displacement (which is along the y direction) is at right angles to the direction of propagation of the wave, we have what is known as a *transverse wave*. Now, if we take a snapshot of the string at $t = 0$ and at a slightly later time Δt, the snapshots will look like those shown in Fig. 2.2*a*; the figure shows that the disturbances have identical shapes except for the fact that one is displaced from the other by a distance $v\Delta t$, where v represents the speed of the disturbance. Such a propagation of a disturbance without a change in form is characteristic of a wave. Now, at $x = 0$, we have

$$y(x = 0, t) = a \sin \omega t \tag{2.3}$$

Fig. 2.2*b*, and each point on the string vibrates with the same frequency ν, and therefore if T represents the time taken to complete one vibration, it is simply the inverse of the frequency:

$$T = \frac{1}{\nu} \tag{2.4}$$

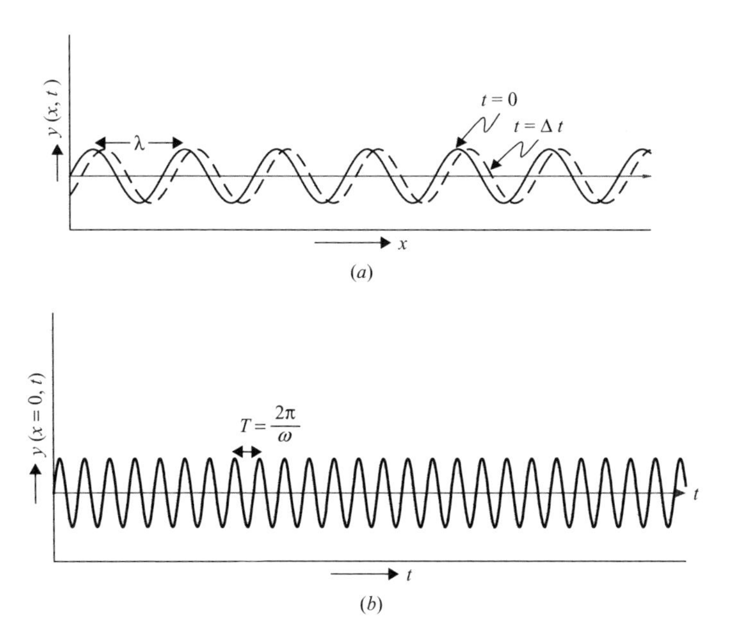

FIGURE 2.2 (*a*) Displacement of a string at $t = 0$ and at $t = \Delta t$, respectively, when a sinusoidal wave is propagating in the $+x$ direction; (*b*) time variation of the displacement at $x = 0$ when a sinusoidal wave is propagating in the $+x$ direction. At $x = \Delta x$, the time variation of the displacement will be slightly displaced to the right.

It is interesting to note that each point of the string moves up and down with the same frequency ν as that of our hand, and the work we do in generating the wave is carried by the wave, which is felt by the person holding the other end of the string. Indeed, all waves carry energy.

Referring back to Fig. 2.2*a*, we note that the two curves are the snapshots of the string at two instants of time. It can be seen from the figure that at a particular instant, any two points separated by a distance λ (or multiples of it) have identical displacements. This distance is known as the *wavelength* of the wave. Further, the shape of the string at the instant Δt is identical to its shape at $t = 0$, except for the fact that the entire disturbance has traveled through a certain distance. If v represents the speed of the wave, this distance is simple $v\ \Delta t$. Indeed, in one period (i.e., in time T) the wave travels a distance equal to λ. Thus, the wavelength of the wave is nothing but the product of the velocity and time period of the wave:

$$\lambda = vT \tag{2.5}$$

which implies that the velocity of the wave is the product of the wavelength and the frequency of the wave:

$$v = \nu\lambda \tag{2.6}$$

Unlike the waves on a string, which are *mechanical waves*, light waves are characterized by changing electric and magnetic fields and are referred to as *electromagnetic waves*. In the case of light waves, a changing magnetic field produces a time- and space-varying electric field, and the changing electric field in turn produces a time- and space-varying magnetic field; this results in the propagation of the electromagnetic wave even in free space. The electric and magnetic fields associated with a light wave can be described by the equations:

$$\mathbf{E} = \hat{\mathbf{y}}\, E_0 \cos(\omega t - kx) \tag{2.7}$$

$$\mathbf{H} = \hat{\mathbf{z}}\, H_0 \cos(\omega t - kx) \tag{2.8}$$

where E_0 represents the amplitude of the electric field (which is in the y direction) and H_0 represents the amplitude of the magnetic field (which is in the z direction); $\hat{\mathbf{y}}$ and $\hat{\mathbf{z}}$ are unit vectors along the y and z directions, respectively. Equation (2.7) describes a y-polarized electromagnetic wave propagating in the x direction. Further, $\omega/k = v$ is the velocity of the electromagnetic waves, and in free space $v = c \approx 3 \times 10^8$ m/s. In contrast, sound waves need a medium to propagate since they are formed by mechanical strains produced in the medium in which they propagate.

For propagation along the x direction one could also have an electromagnetic wave whose electric field points along the the z direction while the magnetic field points along the $-y$ direction. The electric field of this wave is perpendicular to the electric field given by Eq. (2.7) and represents a z-polarized wave. The y- and z-polarized waves are the two polarization states of the light wave that can propagate along the x direction.

The intensity of the light wave, which represents the amount of energy crossing a unit area perpendicular to the direction of propagation in a unit time. The intensity I and the peak electric field E_0 of an electromagnetic wave are related to each other through the equation:

$$I = \frac{n}{2c\mu_0} E_0^2 \tag{2.9}$$

where n is the refractive index of the medium through which the wave is propagating and μ_0 is a constant with the value $4\pi \times 10^{-7}$ SI units.

As an example, we can consider a light beam with a cross-sectional diameter of 2 mm propagating through free space. If the power carried by the beam is 1 W, the intensity of the field is 3×10^5 W/m^2, and the electric field associated with this wave would be about 15,500 V/m.

We mention here that a low-powered ($\approx$2 mW) diffraction-limited laser beam incident on the eye gets focused on a very small spot and can produce an intensity of about 10^8 W/m^2 at the retina; this could indeed damage the retina. On the other hand, when we look at a 20-W bulb at a distance of about 5 m from the eye, the eye produces an image of the bulb on the retina, and this would produce an intensity of only about 10 W/m^2 on the retina of the eye. Thus, whereas it is quite safe to look at a 20-W bulb, it is very dangerous to look directly into a 2-mW laser beam. Indeed, because a laser beam can be focused to very narrow areas, it has found important applications in such areas as eye surgery and laser cutting.

It is of interest here to note that if we look directly at the sun, the power density in the image formed is about 30 kW/m^2. This follows from the fact that on Earth, about 1.35 kW of solar energy is incident (normally) on an area of 1 m^2. Thus, the energy entering the eye is about 4 mW. Since the sun subtends about 0.5° on Earth, the radius of the image of the sun (on the retina) is about 2×10^{-4} m. Therefore, if we are looking directly at the sun, the power density in the image formed is about 30 kW/m^2. The corresponding electric field is about 4700 V/m. *Never look into the sun; your retina would be damaged: not only because of the high intensities but also because of the high level of ultraviolet light in sunlight.*

Lasers can generate extremely high powers, and since they can also be focused to very small areas, it is possible to generate extremely high intensity values. At currently achievable intensities such as 10^{21} W/m^2, the electric fields are so high that electrons can get accelerated to relativistic velocities (velocities approaching that of light), leading to very interesting effects. Apart from scientific investigations of extreme conditions, continuous-wave lasers having power levels of about 10^5 W, and pulsed lasers having a total energy of about 50,000 J have many applications (e.g., welding, cutting, laser fusion, Star Wars).

The wave represented by Eq. (2.7) represents a monochromatic wave since it has only one frequency component, represented by ω. We shall see in Chapter 3 that when a wave of the type represented by Eq. (2.7) is modulated in amplitude or frequency according to a signal to be transmitted, this process leads to a wave which then contains many frequency components. In a light pulse the amplitude of the electric field varies with time (Fig. 2.3), and such a field has many frequency components. The frequency spectrum of the pulse is related inversely to the pulse width in time. Thus, a shorter pulse would have a broader spectrum, and conversely, a broader pulse would have a narrower spectrum. The spectrum occupied by a pulse is an important feature and finally determines the information capacity of the fiber optic system.

There exists a wide and continuous variation in the frequency (and wavelength) of electromagnetic waves. The electromagnetic spectrum is shown in Fig. 2.4. Radio waves correspond to wavelength in the range 10 to 1000 m, whereas the wavelength of x-rays are in the region of angstroms (1 Å $= 10^{-10}$ m). The ranges of the wavelengths of various types of electromagnetic waves are shown in Fig. 2.4, and as can be seen, the visible region ($0.4\ \mu\text{m} < \lambda < 0.7\ \mu\text{m}$) occupies a very small portion of the spectrum. Although the range noted above represents the visible range for humans, there are animals and insects whose sensation can extend to regions not visible to humans. For example, pit vipers can sense infrared radiation (heat radiation), and bees are sensitive to ultraviolet radiation, which helps them locate sources of honey. Special cameras that convert infrared radiation to visible light help humans to see objects even in the dark.

The methods of production of various types of electromagnetic waves are different; for example, x-rays are usually produced by the sudden stopping or deflection of electrons, whereas radio waves may be produced by oscillating charges on an antenna.

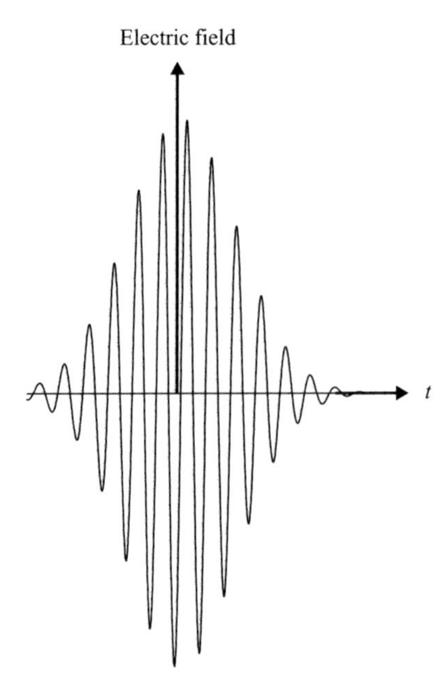

FIGURE 2.3 Optical pulse; the oscillatory portion is due to the high frequency of the pulse, and the envelope is the pulse shape.

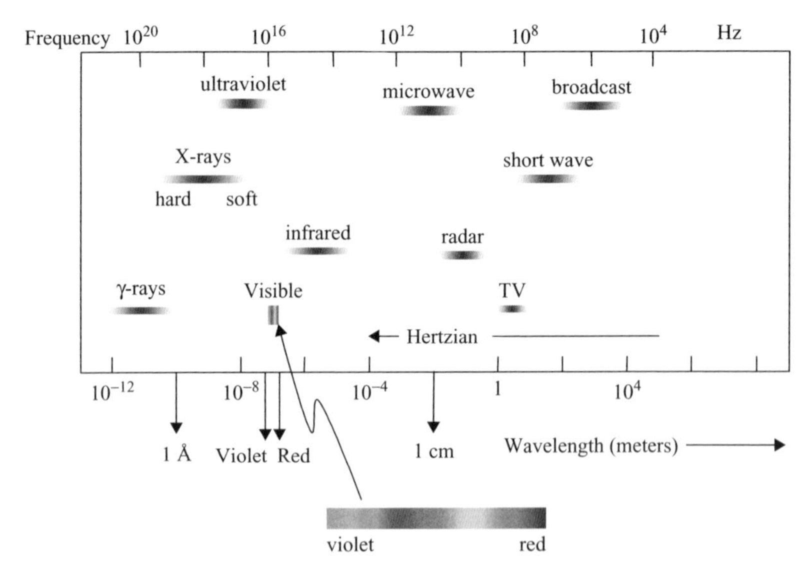

FIGURE 2.4 Electromagnetic spectrum.

However, all electromagnetic waves propagate with the same speed in vacuum, and this speed is denoted by c and is equal to 299,792.458 km/s. This value is usually approximated by 300,000 km/s. Thus, whether it is ultraviolet light or infrared light or radio waves, they all travel with an identical velocity in vacuum.

Knowing the wavelength and the velocity, one can calculate the corresponding frequencies. Thus, yellow light corresponding to a wavelength of 600 nm would have a frequency of 500,000 GHz, where 1 GHz (1 gigahertz) $= 10^9$ Hz ($=1$ billion vibrations per second), so the frequency is 0.5 million GHz (i.e., the electric and magnetic fields oscillate 5 hundred thousand billion times per second!). Compare this with audible sound waves at, say, a frequency of 5 kilohertz, where the vibrations take place only 5000 times per second. On the other hand, for $\lambda = 30$ m (shortwave radio broadcast), the corresponding frequency is 10 megahertz (i.e., oscillations take place 10 million vibrations per second).

According to the theory of relativity, the highest velocity that any wave or object can have is the velocity of light in free space. This velocity is so high that in 1 second, light can travel about 7.5 times around the Earth, and it takes only about 8 minutes for light from the sun to reach us. Similarly, radio signals from the probe that has landed recently on Titan (one of the moons of Saturn) will take about 1.2 hours to reach the radio station on Earth. If we look at a star that is, say, 10 light-years away (i.e., light takes 10 years to reach us from that star), the light that reaches us right now from the star started its journey 10 years ago, and what we are witnessing right now happened 10 years ago!

2.3 REFRACTIVE INDEX

Light waves travel at a slightly slower speed when propagating through a medium such as glass or water. The ratio of the speed of light in vacuum to that in the medium, known as the *refractive index* of the medium, is usually denoted by the symbol n:

$$n = \frac{c}{v} \tag{2.10}$$

where c ($\approx 3 \times 10^8$ m/s) is the speed of light in free space and v represents the velocity of light in that medium. For example,

$$n \approx \begin{cases} 1.5 & \text{for glass} \\ \frac{4}{3} & \text{for water} \end{cases}$$

Thus, in glass, the speed of light $\approx$200,000 km/s, and in water, the speed of light $\approx$225,000 km/s.

When a ray of light is incident at the interface of two media (e.g., air and glass), it undergoes partial reflection and partial refraction as shown in Fig. 2.5*a*. The dotted line represents the normal to the surface. The angles ϕ_1, ϕ_2, and ϕ_r represent the angles that the incident ray, refracted ray, and reflected ray make with the normal.

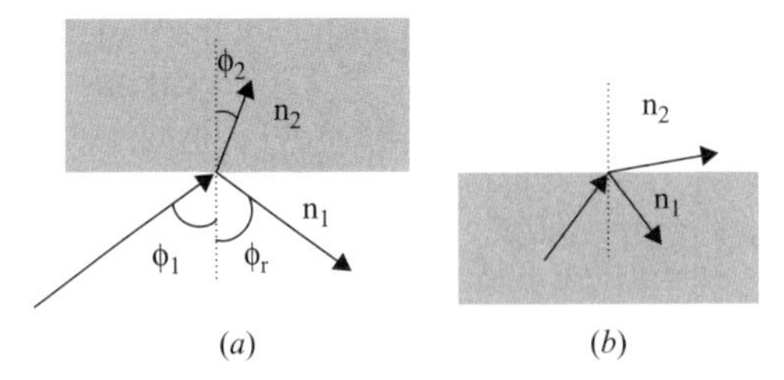

FIGURE 2.5 (*a*) Ray of light incident on a denser medium ($n_2 > n_1$); (*b*) ray incident on a rarer medium ($n_2 < n_1$).

According to *Snell's law*,

$$n_1 \sin \phi_1 = n_2 \sin \phi_2 \quad \text{and} \quad \phi_r = \phi_1 \tag{2.11}$$

Further, the incident ray, reflected ray, and the refracted ray lie in the same plane. In Fig. 2.5*a*, since $n_2 > n_1$, we must have (from Snell's law) $\phi_2 < \phi_1$ (i.e., the ray will bend towards the normal). On the other hand, if a ray is incident at the interface of a rarer medium ($n_2 < n_1$), the ray will bend away from the normal as shown in Fig. 2.5*b*.

Example 2.1 For the air–glass interface, $n_1 = 1.0$, $n_2 = 1.5$ and if $\phi_1 = 45°$, then $\phi_2 \simeq 28°$ (Fig. 2.6*a*). Similarly, for the air–water interface, $n_1 = 1.0$, $n_2 = 1.33$ and if $\phi_1 = 45°$, then $\phi_2 \simeq 32°$ (Fig. 2.6*b*).

The path of rays is reversible; that is, if a light ray (passing through water) is incident on air, the ray will bend away from the normal. Figure 2.7 shows exactly the reverse of the situation in Fig. 2.5*b*, where the ray is incident from water and refracts into air. It is because of this refraction that when we look at a fish (which is inside the

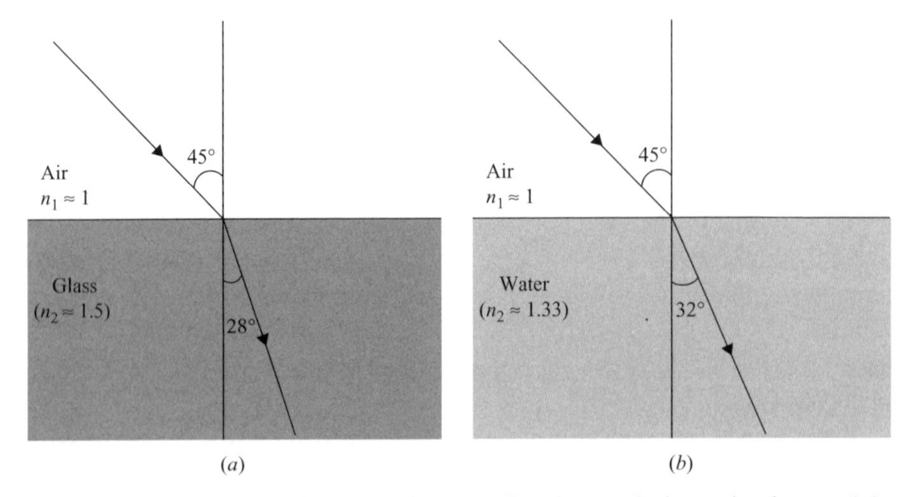

FIGURE 2.6 For a ray incident on a denser medium ($n_2 > n_1$), the ray bends toward the normal and the angle of refraction is less than the angle of incidence: (*a*) for the air–glass interface, for $\phi_1 = 45°$, $\phi_2 \approx 28$; (*b*) for the air–water interface, for $\phi_1 = 45°$, $\phi_2 \approx 32°$.

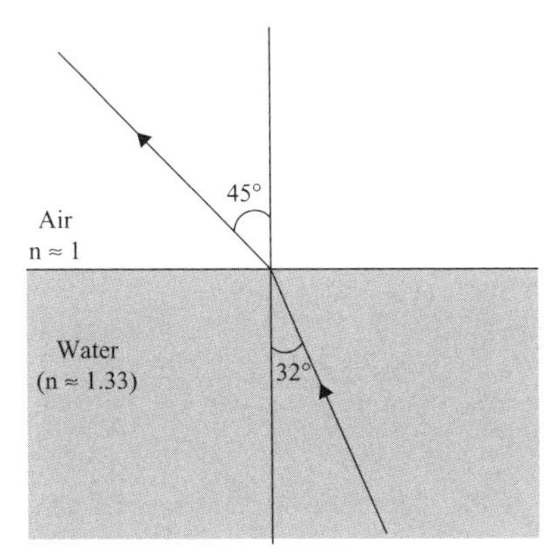

FIGURE 2.7 For a ray incident on a rarer medium, the ray bends away from the normal, and the angle of refraction is greater than the angle of incidence. For the air–glass interface, for $\phi_1 = 32^\circ$, the angle of refraction is 45°.

water) from outside, the fish appears closer to the surface, as shown in Fig. 2.8; or when we view a pencil partially dipped in a glass of water, it seems to be bent. The outside world as seen by a fish is quite different, due to the phenomenon of refraction. The entire horizon (cone of 180°) is condensed into a cone of approximately 96° (Fig. 2.9). Thus, the apparent position of objects seen by the fish is different from the actual position. Surprisingly, some fishes seem to have learned about refraction since they take into account the apparent position of the prey (brought about by refraction of light) before striking them. The archer fish presents a very interesting example, since to catch a prey it squirts a jet of water out of its mouth onto its victim sitting on a plant outside water on land about 2 m away and knocks it down.

As mentioned above, if a ray is incident at the interface of a rarer medium (i.e., a medium with lower refractive index), the ray will bend away from the normal (Fig. 2.6). As we increase the angle of incidence, the angle of refraction will become larger.

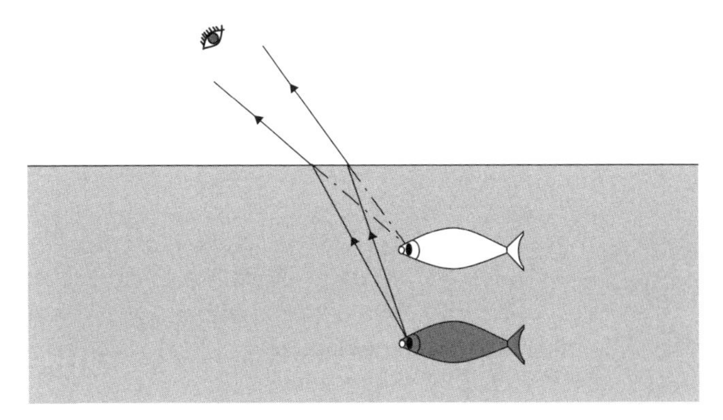

FIGURE 2.8 When we look from outside at a fish (which is inside the water), because of the refraction of light, the fish appears closer to the surface.

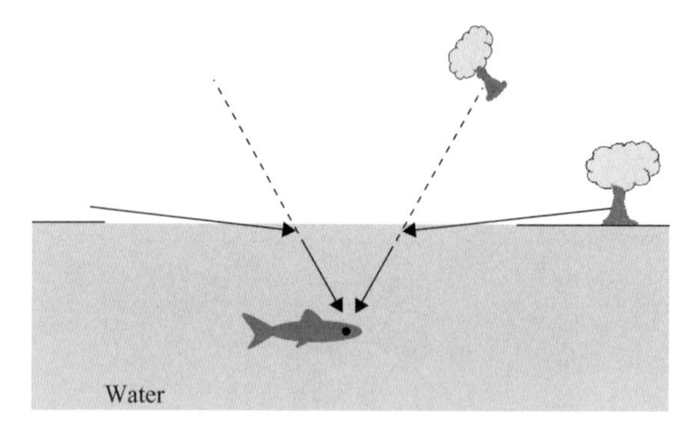

FIGURE 2.9 The world as seen by a fish.

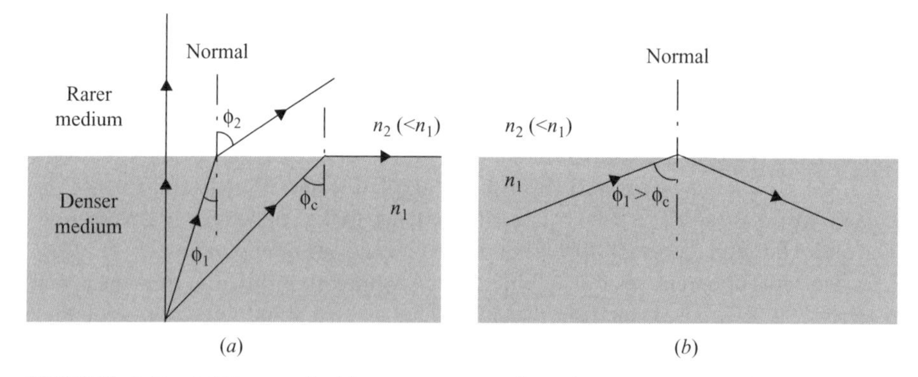

FIGURE 2.10 (*a*) For a ray incident on a rarer medium, the ray bends away from the normal and the angle of refraction is greater than the angle of incidence. For the air–glass interface, for $\phi_1 = \phi_c \approx 41.8°$, the angle of refraction is 90°; this is the critical angle. (*b*) If the angle of incidence is greater than critical angle, it will undergo total internal reflection.

The angle of incidence for which the angle of refraction is 90° is known as the *critical angle* (Fig. 2.10*a*) and is denoted by ϕ_c. Thus, $\phi_2 = 90°$ when

$$\phi_1 = \phi_c = \sin^{-1}\frac{n_2}{n_1} \tag{2.12}$$

Example 2.2 For the glass–air interface, $n_1 = 1.5$, $n_2 = 1.0$ and the critical angle is given by

$$\phi_c = \sin^{-1}\frac{1}{1.5} \approx 41.8°$$

On the other hand, for the glass–water interface, $n_1 = 1.5$, $n_2 = \frac{4}{3}$, and

$$\phi_c = \sin^{-1}\frac{1.33}{1.5} \approx 62.7°$$

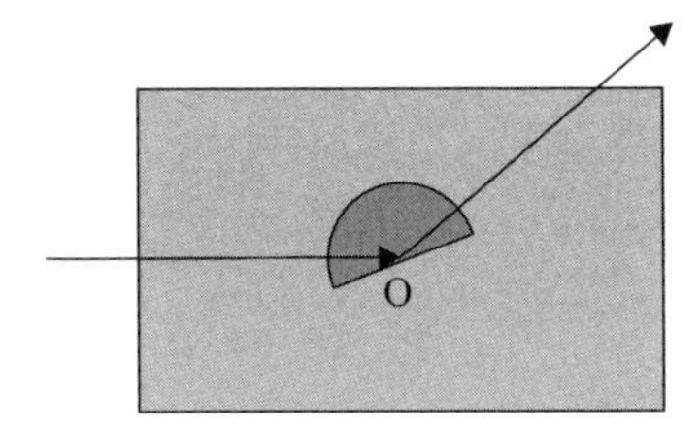

FIGURE 2.11 Simple laboratory experiment to demonstrate the phenomenon of total internal reflection.

If a ray of light is incident at an angle of incidence greater than the critical angle, there is no refracted ray and all the incident energy is reflected back Fig. 2.10*b*. This is known as *total internal reflection*, a subject of great practical importance.

The phenomenon of *total internal reflection* (TIR) can be very easily demonstrated through a simple experiment, as shown in Fig. 2.11. A thick semicircular glass disk is immersed in a glass vessel filled with water. A laser beam from a helium–neon (He–Ne) laser or simply a laser pointer is directed toward the center of the semicircular disk so that it is incident normally on the glass surface and goes undeviated, as shown in the figure. The angle of incidence (at the glass–water interface) is increased by rotating the glass disk about point *O*; eventually, when the angle of incidence exceeds the critical angle, the laser beam undergoes total internal reflection, which can be seen clearly when viewed from the top. If one puts in a drop of ink in water, the light path becomes very beautiful to look at! The experiment is very simple and we urge the reader to carry it out using a laser pointer.

Although the phenomenon of TIR has been known for hundreds of years, the first experimental demonstration of light guidance through total internal reflection was carried out by sending a light beam through a water jet; this was first demonstrated in 1843 by Daniel Colladon and indepentenly by Jacques Babinet. A schematic of this demonstration is shown in Fig. 2.12; light undergoes total internal reflection at the water–air interface and travels along the curved path of water emanating from

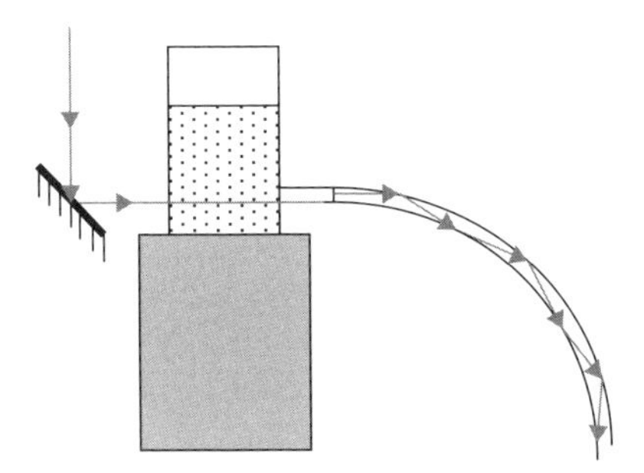

FIGURE 2.12 Light guidance through a water jet, demonstrating the phenomenon of total internal reflection; this was first demonstrated by Daniel Colladon in 1841.

an illuminated vessel. We should mention here that John Tyndall is usually credited with the first demonstration of light guidance in water jets; however, he did not demonstrate light guiding in water jets until 1855, duplicating but not acknowledging his predecessors. For a very nice historical survey, we refer the reader to the book by Hecht (1999).

2.4 GRADED-INDEX MEDIA

A homogeneous medium is one in which the refractive index of the medium is the same throughout. The media we discussed above are homogeneous media, and in such media, light rays travel along straight lines. Graded-index or inhomogeneous media are media in which the refractive index varies with position (i.e., the refractive index of the medium is different at different points).

A very interesting example of a medium with varying refractive index is found in the formation of mirages. If you recall, when we look along the ground or the road, on a hot day we can see a mirage: apparent reflection of objects from the ground, giving a feeling of the presence of water. This happens due to the specific paths that light rays emanating from objects take while propagating through the air column. The ground, being hot, heats up the air in contact with it while the air at a height is cooler. Thus, the temperature of air decreases as we go above the ground. The refractive index of a gas depends on the temperature, and this temperature variation also leads to an increase in the refractive index as we move up from the ground. Now, imagine a light ray emanating from an object as shown in Fig. 2.13. A light ray from the tree propagating downward encounters media of lower refractive indices. We saw earlier that when a light ray propagates from a denser medium to a rarer medium, it bends away from the normal. Thus, as the light ray propagates toward the ground, it bends continuously away from the normal (which in this case is vertical). If it becomes horizontal before hitting the ground, it turns back and now starts to go upward (Fig. 2.13); rays hitting the ground are, of course, lost. A specific curved ray

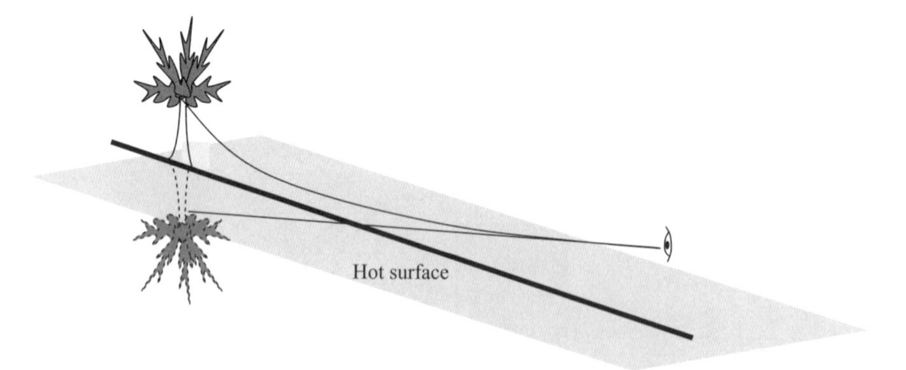

FIGURE 2.13 Formation of a mirage due to the increase in the refractive index of air with height brought about by the decrease of temperature with height.

would reach the eye of the observer. Now, the upper portion of the object is in a region of almost constant refractive index, and rays at that height would almost travel along a straight line and would also reach the eye. Hence, in this case the observer sees the object due to rays reaching the eye in almost straight-line paths as well as a virtual image of the object due to rays appearing from the direction of the ground. These rays will give the feeling of reflection from the surface of the ground and form the mirage.

2.5 DISPERSION

The velocity of propagation of a light wave in any medium, and hence the refractive index of the medium, depend slightly on the wavelength of the propagating light wave. Normally, as the wavelength increases, the refractive index of the medium decreases. Because of the slight variation of refractive index with wavelength, if light containing many wavelengths (e.g., white light) is incident at an interface between two media, the angle of refraction will be different for different wavelengths. The dependence of the refractive index on wavelength leads to what is known as *dispersion*.

In Fig. 2.14 we see a narrow pencil of a white light beam incident on a prism. Since the refractive index of glass depends on the wavelength, the angle of refraction will be different for different colors, and the incident white light will *disperse* into its constituent colors—the dispersion will become greater at the second surface of the prism (Fig. 2.14).

The phenomenon of dispersion is responsible for the formation of rainbows since different wavelengths present in the sunlight refract into different angles as they enter the water droplets present in the atmosphere. After reflection from the water–air interface they refract out and are seen by our eyes (Fig. 2.15). Sometimes we can

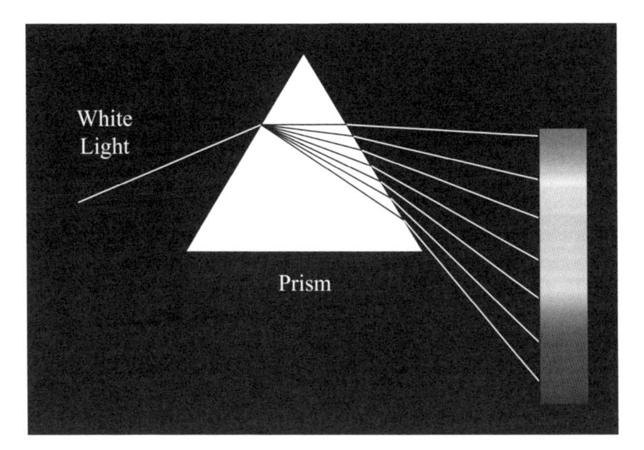

FIGURE 2.14 Dispersion of white light as it passes through a prism. Red color appears at the top and violet color at the bottom.

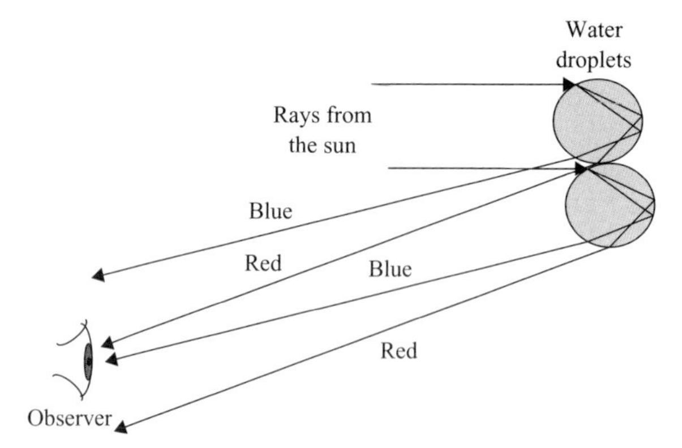

FIGURE 2.15 Rainbows are formed due to dispersion of light when they refract in water droplets.

observe a second rainbow outside the primary one (Fig. 2.16). If you notice carefully, the colors of the secondary rainbow are in reverse order to that of the primary rainbow, and it is also not as bright as the primary rainbow. The secondary rainbow is formed by further reflection and refraction within the water droplets.

In Chapter 6 we will see that dispersion in silica (which is the base constituent of glass), which is the primary component of optical fibers, is responsible for the broadening of optical pulses as they propagate through an optical fiber.

FIGURE 2.16 The inner one is called the *primary rainbow* and the outer one is called the *secondary rainbow*.

CHAPTER 3

Carrier Wave Communication

3.1 INTRODUCTION

Communication using electromagnetic waves are today the most reliable, economical, and fastest way of communicating information between points. In any communication system, the information to be transmitted is generated at a source; gets transmitted through a channel such as the atmosphere in a radio broadcast, or electrical lines in a telephone or wireless network in mobile communication or optical fibers in a fiber optic communication system; and finally, reaches a receiver, which is the destination. Usually, the channel through which information propagates introduces loss in the signal and also distorts it to a certain extent. For a communication system to be reliable, the channel must introduce minimal distortion to the signal. There should also be very little noise added by the channel so that the information can be retrieved without significant errors.

The electrical signals produced by various sources, such as the telephone, computer, or video, are not always suitable for transmission directly as such through the channel. These signals are made to modulate a high-frequency electromagnetic wave such as a radio wave, microwave, or light wave, and it is this modulated electromagnetic wave that carries the information. Such a communication systems is referred to as *carrier wave communication*.

There are different ways of modulating an electromagnetic wave in accordance with a given signal. The modulation can be either analog or digital. In *analog modulation*, the amplitude, phase, or frequency of the carrier wave is changed in accordance with the signal amplitude; in *digital modulation*, the analog signal is first converted into a digital signal consisting of 1's and 0's, which is then used to modulate the carrier. In the following we discuss these schemes.

3.2 ANALOG MODULATION

In analog modulation some characteristic of the carrier wave (amplitude, phase, or frequency) is modulated in accordance with the signal; the characteristic can take

---

Fiber Optic Essentials, By K. Thyagarajan and Ajoy Ghatak

values continuously within a range. Since the carrier wave is a sinusoidal wave, we can represent the carrier wave by the equation:

$$V(t) = V_0 \sin(\omega t - \phi) \tag{3.1}$$

where V_0 is a constant and ω represents the carrier frequency; ϕ is an arbitrary phase. Here V represents either the voltage or the electric field of the electromagnetic wave. Amplitude, phase, and frequency modulations correspond to modulating the amplitude, phase, and frequency of the carrier wave.

Amplitude Modulation

In *amplitude modulation* (AM), the amplitude of the carrier wave is modulated in accordance with the signal to be sent. Thus, we can write for an amplitude-modulated wave,

$$V(t) = V_0[1 + m(t)] \sin(\omega t - \phi) \tag{3.2}$$

where $m(t)$ represents the time-varying signal to be transmitted. As the signal amplitude changes, the amplitude of the modulated wave changes and thus the modulated signal carries the information.

As an example, if we consider the signal to be another sine wave with frequency Ω ($\ll \omega$), we have the modulated wave as

$$V(t) = V_0[1 + a \sin \Omega t] \sin(\omega t - \phi) \tag{3.3}$$

where a is a constant. Expanding the term in brackets and using the formulas for the product of sine functions, we have

$$V(t) = V_0\{\sin(\omega t - \phi) + \tfrac{1}{2}a \cos[(\omega - \Omega)t - \phi] - \tfrac{1}{2}a \cos[(\omega + \Omega)t - \phi]\} \tag{3.4}$$

Hence, the modulated wave now contains three frequencies, ω, $\omega + \Omega$, and $\omega - \Omega$: the carrier, upper sideband, and lower sideband frequencies. Thus, amplitude-modulating a carrier wave by a sinusoidal wave generates two sidebands. Since any general time-varying function can be analyzed in terms of sinusoidal functions, amplitude modulation of the carrier wave would result in the generation of an upper and a lower sideband. If the maximum frequency of the signal is $\Omega_{\max}$, the upper sideband would lie between ω and $\omega + \Omega_{\max}$, the lower sideband from $\omega - \Omega_{\max}$ to ω. Both the sidebands contain information of the entire signal.

Figure 3.1*a* shows a signal to be transmitted and Fig. 3.1*b* shows the carrier wave; notice that the frequency of the carrier wave is much larger than the frequencies contained in the signal. Figure 3.1*c* shows the amplitude-modulated carrier wave. The signal now rides on the carrier as its amplitude modulation. At the receiver, the modulated carrier is demodulated and the signal can be retrieved.

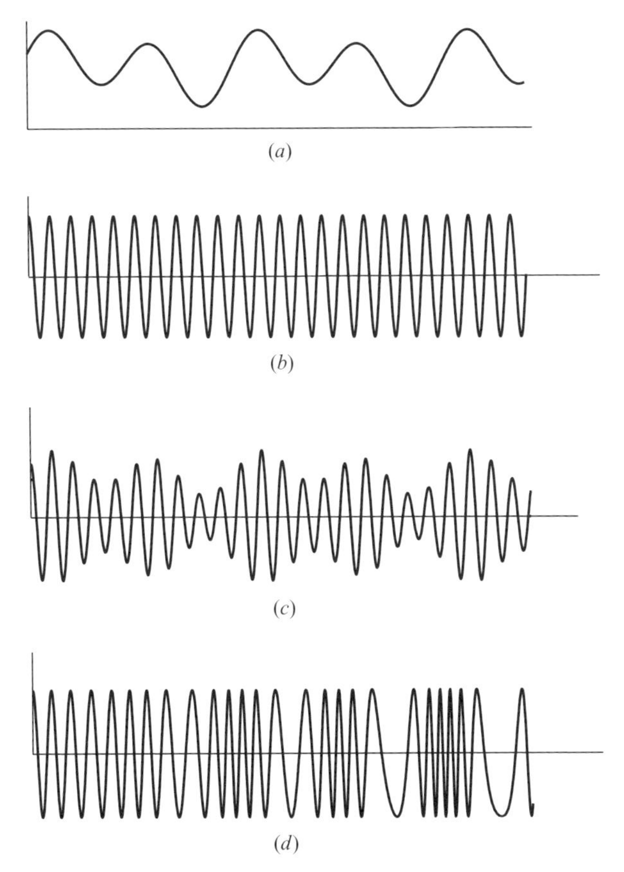

FIGURE 3.1 (*a*) Signal to be transmitted; (*b*) carrier wave using the signal to be communicated; (*c*) amplitude-modulated carrier wave; (*d*) frequency-modulated carrier wave.

As an example, let us assume that we wish to transmit speech. We first note that for speech to be intelligible, it is sufficient to send signal content up to a frequency of 4000 Hz. Thus, the electrical signal that comes out of the microphone into which the person speaks can be restricted to a frequency of 4000 Hz. In comparison, for sending high-fidelity music, the upper frequency is 20,000 Hz, which is also the limit of frequency of human hearing. The electrical signal from the microphone could look like the signal shown in Fig. 3.1*a*, and if we are considering radio transmission, the radio wave on which the information rides will look like the wave shown in Fig. 3.1*b*. The amplitude-modulated radio wave would then be like the one shown in Fig. 3.1*c*. It is this modulated wave that is broadcast through open space (which is the channel), and at the receiver (your radio set) it is demodulated, the signal is retrieved, and you hear the speech.

An obvious question that arises is: How can one send more than one signal simultaneously through the same channel: for example, the atmosphere in radio communication? To understand this we first note that the carrier wave shown in Fig. 3.1*b* is at a single frequency, whereas the amplitude-modulated signal shown in Fig. 3.1*c* has a spectrum (i.e., it has a range of frequencies). Thus, if the signal occupies the frequencies up to 4000 Hz, and if the carrier wave frequency is 1,000,000 Hz, the amplitude-modulated wave has frequencies lying between 996,000 and 1,004,000 Hz (sum and difference of carrier frequency and the maximum signal frequency). The information contained in the frequency range 996,000 to 1,000,000 Hz is the lower sideband, and the information contained in the frequency range 1,000,000 to 1,004,000 Hz is the upper sideband, and the bands together contain all the information. Hence, it is sufficient to send only one of the sidebands (e.g., the components lying between 1,000,000 and 1,004,000 Hz) in order for the receiver to retrieve the signal; this is referred to as upper sideband transmission. Hence, we see that to send one speech signal we need to reserve the frequencies lying between 1,000,000 and 1,004,000 Hz, a band of 4000 Hz.

Now, to send another speech signal, we can choose a radio wave of frequency 1,004,000 Hz and send the modulated wave lying in the frequency band 1,004,000 and 1,008,000 Hz. These frequencies lie outside the range of the frequencies of the first signal and hence will not interfere with that signal: similarly for more and more speech signals. Thus, if we can use carrier frequencies over a range of, say, 1,000,000 to 3,000,000 Hz, we can send 2,000,000/4000 = 500 speech signals simultaneously. This also makes it clear that the larger the range of frequencies of the carrier wave, the larger the number of channels that can be sent simultaneously. The range of frequencies available increases with the frequency of the carrier wave, and this is the reason why light waves that have frequencies much higher than radio waves or microwaves can transmit much more information.

Frequency Modulation

In frequency modulation (FM), instead of modulating the amplitude of the carrier wave, its frequency is changed in accordance with the signal, as shown in Fig. 3.1*d*. In this case, information is contained in the form of the frequency of the signal. For the case of the frequency-modulated signal, instead of Eq. (3.3) we would have

$$V(t) = V_0 \sin\{\omega[1 + am(t)]t - \phi\} \tag{3.5}$$

As can be seen, in this case the amplitude of the wave remains constant while the frequency changes with time in accordance with the signal represented by $m(t)$. Equation (3.5) represents a wave that does not have just one frequency but many frequency components. The frequency spectrum in this case is not as simple as in the case of amplitude modulation. It can be shown that unlike the amplitude-modulation case, where the amplitude-modulated signal had a narrow upper and a narrow lower sideband, in the case of frequency modulation the modulated signal contains many more frequency components. Hence, an FM signal requires a much larger bandwidth to transmit than an AM signal. Thus, for a given range of carrier frequencies, the number of independent channels that can be sent using frequency modulation would be smaller. To accommodate more channels, the carrier frequencies used in frequency modulation are much higher and fall in the range 30 million to 300 million Hz (30 to 300 MHz). Since the information is coded into the frequency of the carrier wave, the frequency-modulated waves are less susceptible to noise, and this is quite apparent while listening to an AM radio broadcast (medium- or shortwave channels) or an FM radio broadcast.

In the foregoing methods, simultaneous transmission of different independent signals is accomplished by reserving different carrier frequencies for different signals. This method is referred to as *frequency-division multiplexing*. All the signals are propagating simultaneously through the transmission medium and the receiver can pick up any of the signals by filtering only the frequency band of interest to it (i.e., tune into the required signal).

3.3 DIGITAL MODULATION

The modulation scheme used in optical fiber communication is called *digital modulation*. The digital modulation scheme is based on the fact that an analog signal satisfying certain criteria can be represented by a digital signal. There is a theorem, called the *sampling theorem*, according to which a signal that is limited by a maximum frequency (also referred to as a *bandlimited signal*), that is, a signal that has no frequency component above a certain frequency, say, ν_m, is determined uniquely by its values at uniform time intervals spaced less than $1/2\ \nu_m$. Thus, if we consider speech that has frequencies below 4000 Hz, the analog speech signal (like the one shown in Fig. 3.1*a*) can be represented uniquely by specifying the values of the signal at time intervals of less than 1/8000 s. Thus, if we sample the speech signal at 8000 times per second, and if we are given the values of the signal at these times, we can uniquely determine the original analog speech signal even though we are not told the value of the function at intermediate points! Figure 3.2 represents this fact; Fig. 3.2*a* shows the same signal as Fig. 3.1*a*, and Fig. 3.2*b* represents the sampled values at

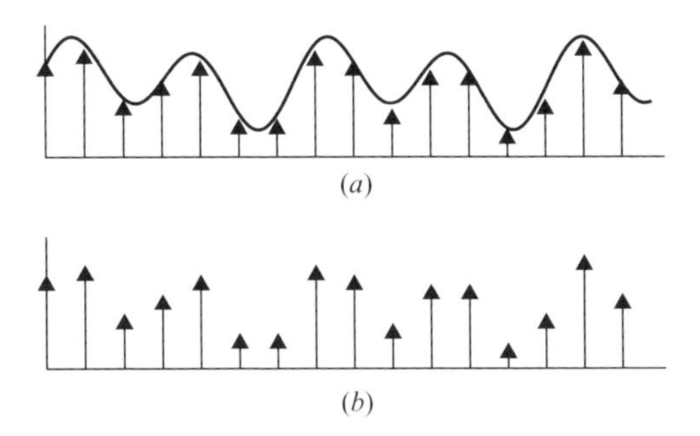

FIGURE 3.2 (*a*) Sampling of the given analog signal; (*b*) sampled values of the signal. Note that even though the values of the signal between the samples is not specified, these can be determined uniquely from the sampled values.

time intervals of 1/8000 s. Thus, instead of sending the analog signal, it is sufficient to send the values of the signal at specific times, and this is sufficient to determine the signal.

Instead of sending pulses of different amplitudes corresponding to different sampled values, it is usual first to convert the various pulse amplitudes into a binary signal that will consist of only two values of amplitudes: *high amplitude*, referred to as 1, and *low amplitude* (usually zero), referred to as 0. At this stage it may be worthwhile to look at an example of what a binary system can perform. This is an example taken from the book *The Road Ahead* by Bill Gates (1996).

Let us assume that we need to illuminate a room with 250 W of light and we wish the illumination to be variable and adjustable from 0 to 250 W in steps of 1 W. To achieve this we can have one bulb of 250-W power connected to a dimmer and adjust the dimmer to achieve any value that one wishes. In this technique it is very difficult to set the bulb to exactly the same illumination repeatedly, since positioning the knob to exactly the same position is not really feasible. Also, if we wish to have another room illuminated with exactly the same illumination, it would not be very easy since we are again restricted by the ability to position two knobs corresponding to exactly the same position.

A very interesting way to achieve exactly the same illumination repeatedly is to use eight bulbs with power levels of 1, 2, 4, 8, 16, 32, 64, and 128 W. In this sequence you can notice that the power of each bulb is twice, that of the preceding one. Now it is possible to generate using these bulbs any illumination between 0 and 250 W (in fact, up to 255 W) at intervals of 1 W. Thus, if we wish to generate an illumination corresponding to 100 W, we can switch on the 64-, 32-, and the 4-W bulbs, with all other bulbs switched off. Similarly, to achieve, say, 199 W of illumination, we can switch on only the 128-, 64-, 4-, 2-, and the 1-W bulbs. You can indeed verify that using this scheme, it is possible to combine the various bulbs to achieve any wattage of illumination (at intervals of 1 W) from 0 W (all bulbs switched off) to 255 W (all bulbs switched on).

Now if we refer to a switched-on bulb as 1 and a switched-off bulb as 0 and arrange the positions of on and off bulbs from left to right, with the 128-W bulb being the leftmost and the 1-W bulb the rightmost, we can write the two illuminations, 100 W and 199 W, with the following sequence:

Required Illumination	128 W	64 W	32 W	16 W	8 W	4 W	2 W	1 W
100 W	0	1	1	0	0	1	0	0
199 W	1	1	0	0	0	1	1	1

Thus, the sequences of numbers 01100100 and 11000111 represent the numbers 100 and 199, respectively. Any integer between 0 and 255 can be represented by this sequence of 1's and 0's with eight digits. These sequence of 1's and 0's represent the digital equivalent of the decimal numbers 100 and 199. The former representation, in which integers are specified by 1's and 0's, is called *binary representation* (two digits 1 and 0 are employed), while the conventional representation is referred to as *decimal representation* (10 digits, 0 to 9). If a number greater than 255 has to be represented, we need to take a ninth digit before the digit corresponding to 128 and that would then correspond to the decimal number 256. Computers use the digital language for processing of information, and today the binary representation is all pervasive, as it is used in computer disks, digital video disks, and so on.

Pulse Code Modulation

The most common modulation scheme employed in optical fiber communication is pulse-code modulation. In this, each amplitude of the values sampled is represented by a binary number consisting of eight digits (Fig. 3.3). Since the maximum decimal value with eight digits is 255 ($= 128 + 64 + 32 + 16 + 8 + 4 + 2 + 1$), the maximum amplitude of the signal is restricted to 255 so that all integer values of signal can be represented by a sequence of 1's and 0's.

In pulse-code modulation, the given analog signal is first sampled at an appropriate rate, and then the sample values are converted to binary form. The carrier is then modulated using the binary signal values to generate the modulated signal. The most common scheme employed is called *on–off keying* (OOK). In this scheme every digit 1 is represented by a high-amplitude value of the carrier and every digit 0 by a zero amplitude of the carrier. Figure 3.4 shows the modulated wave corresponding to a binary sequence of eight digits, 10011010. In the scheme shown in Fig. 3.4*a* the amplitude of the carrier does not return to zero when there are two adjacent 1's in the signal. This is referred to as a *non-return-to-zero* (NRZ) *scheme*. There is another scheme, called the *return-to-zero* (RZ) *scheme*, in which the amplitude of the carrier returns to zero even if the adjacent digits are 1 (see Fig. 3.4*b*).

One of the major differences between the NRZ and RZ pulse sequences is the bandwidth requirement. To appreciate this we first note that in the NRZ scheme the fastest changes correspond to alternating sequence of 1's and 0's, whereas in

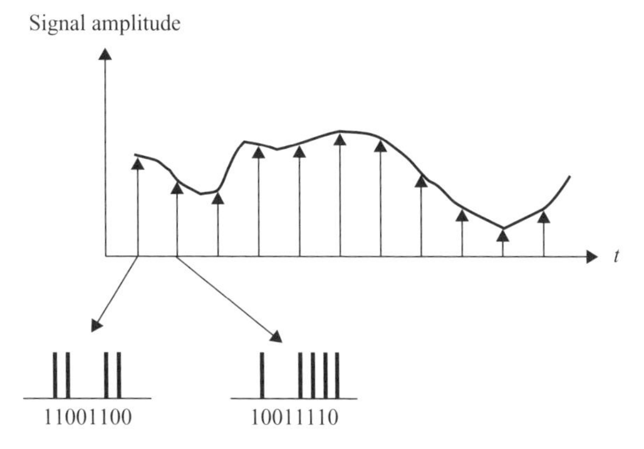

FIGURE 3.3 Binary representation of each of the signal values sampled. Each sampled value is represented by eight binary digits.

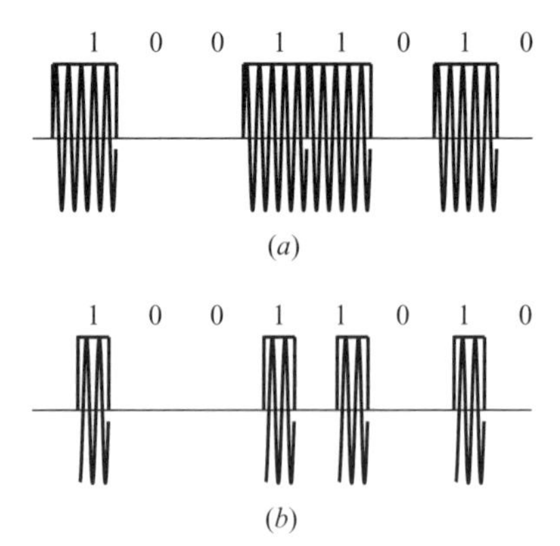

FIGURE 3.4 (*a*) Non-return-to-zero (NRZ) and (*b*) return-to-zero (RZ) schemes.

the case of RZ, a sequence of 1's represents the fastest changes (Fig. 3.5). From Fig. 3.5*a* it can be seen that the fundamental frequency component in the case of NRZ pulse sequence is $1/2T$, whereas in the case of RZ it is $1/T$. Hence for transmission without too much distortion, NRZ would require a bandwidth of at least $1/2T$, while RZ would need a bandwidth of $1/T$. If the bit rate is B, then $B = 1/T$, and thus the bandwidth requirements for NRZ and RZ are given by

$$\Delta f \approx \begin{cases} \frac{B}{2}; & \text{NRZ} \\ B; & \text{RZ} \end{cases} \tag{3.6}$$

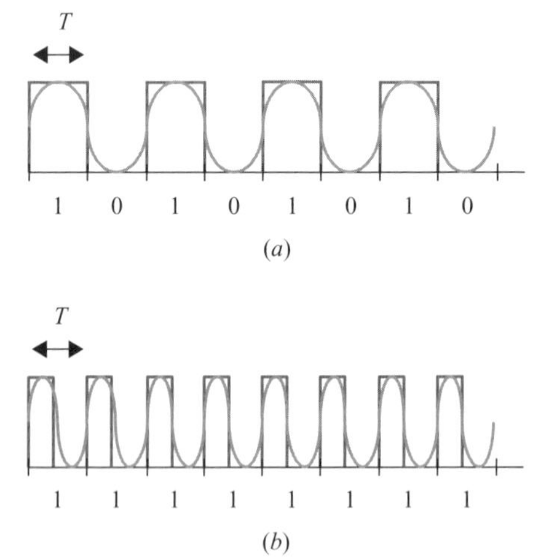

FIGURE 3.5 An alternating sequence of 1's and 0's corresponds to the maximum rate of change in NRZ (*a*) whereas in RZ (*b*) it is a series of 1's. The sinusoidal curves superimposed on the pulses correspond to the fundamental frequency of the pulse sequence.

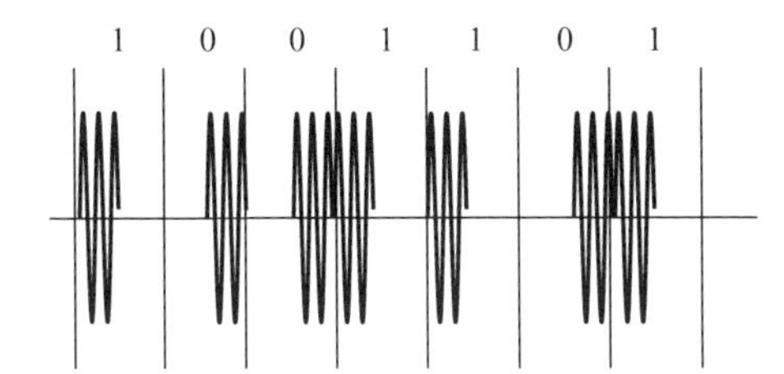

FIGURE 3.6 Manchester coding of the binary signals.

Thus, the bandwidth requirements for RZ are more severe than those for NRZ, which is also expected since pulses in RZ format are narrower than pulses in NRZ format. For example, a 2.5-Gb/s (gigabites per second) system would need 2.5 GHz in RZ format, whereas the same signal would require only 1.25 GHz in the NRZ format. Most fiber optic communication systems today use NRZ schemes for communication.

There are other coding schemes to take care of other issues, such as when a continuous sequence of 1's or 0's appear in a NRZ scheme. In this case the signal would remain constant, and this creates a problem with reception. To overcome this, coding techniques such as Manchester coding have been developed. Figure 3.6 shows a Manchester-coded pulse sequence in which both 1 and 0 contain a transition, with 1 occupying the first half of the time interval and 0 occupying the second half. In this way there would never be a situation in which the input at the receiver would remain constant.

In the amplitude and frequency modulation schemes, different independent signals are allocated different carrier frequencies, leading to frequency-division multiplexing. In digital modulation, the carrier frequency of different channels can be the same, but different independent signals are multiplexed in the time domain. In this concept, shown in Fig. 3.7, two independent signals are sent using the pulse-code modulation scheme, in which the time slots occupied by the two signals are different. Thus, the

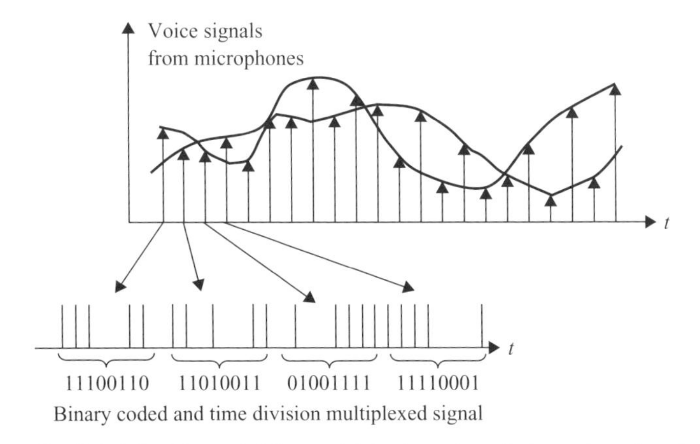

FIGURE 3.7 Time-division multiplexing of two independent signals.

signals now overlap in the frequency domain but are sent at different times, leading to what is referred to as *time-division multiplexing*.

One of the greatest advantages of the pulse-code modulation scheme of communication is that the receiver has to detect only the presence or the absence of a pulse; the presence would correspond to 1 and the absence would correspond to 0. This is unlike the case in amplitude modulation, in which the receiver is supposed to measure the amplitude of the signal precisely. Since detection of the presence or absence of a pulse is much more accurate than actual measurement of the amplitude, communication using pulse-code modulation suffers from much less distortion than do other schemes. Also, for long-distance communication, the signals can be cleansed of noise at regular intervals (at what are called regenerators), and the accumulation of noise can be restricted. Thus, for long-distance, communication pulse-code modulation is much preferred over any analog system scheme.

Since the signals in pulse-code modulation are in the form of pulses, it is very important to study the effect of propagation on these pulses through an optical fiber. If the pulses get distorted with propagation, it is possible that adjacent pulses start to overlap, and this would result in errors in detection of 1's and 0's. This aspect of pulse dispersion is covered in Chapters 6 and 7. In Chapter 8 we also discuss the errors introduced during propagation through a fiber optic communication system.

Bit Rate Required for Speech

We have seen that to transmit speech, we need to send frequencies up to 4000 Hz, and if these have to be represented by digital pulses, we need to sample the signal at least 8000 times per second. Now each of these values sampled would be represented by a binary-digit sequence. In telephony, each value sampled is represented by eight pulses (or 8 bits), and thus the number of pulses per second for each telephone channel would be 64,000, referred to as 64,000 bits per second (b/s) or 64 kilobits per second (kb/s). If we have a communication system capable of transmitting at the rate of 1 billion pulses per second, this would correspond to transmitting (1,000,000,000/64,000) or 15,000 speech signals simultaneously. A system transmitting 1 billion pulses per second is referred to as a 1-gigabit per second (1-Gb/s) system. The higher the bit rate, the larger the capacity of information transmission. Table 3.1 gives the number of bits required for a variety of signals (information content).

TABLE 3.1 Bit Requirements of Some Common Information-Containing Signals

Service	Bit Requirement
1000 words of text	60×10^3 bits
Telephone	64×10^3 b/s
20-volume encyclopedia	3×10^8 bits
Standard TV	100×10^6 b/s
High-definition TV	1.2×10^9 b/s

TABLE 3.2 Hierarchy of Digital Signals in Two Common Data Rates

SDH Signal	SONET Signal	Bit Rate (Mb/s)
STM-0	OC-1	51.840
STM-1	OC-3	155.520
STM-4	OC-12	622.080
STM-16	OC-48	2, 488.320
STM-64	OC-192	9, 953.280
STM-256	OC-768	39, 813.120

Standard Bit Rates

Table 3.2 presents the hierarchy of two common standard bit rates, SDH (Synchronous Digital Hierarchy) and SONET (Synchronous Optical Network), used for data transmission over optical fiber networks. SDH is the international version published by the International Telecommunications Union (ITU), and SONET is the US version of the standard published by the American National Standards Institute (ANSI).

CHAPTER 4

Optical Fiber

4.1 INTRODUCTION

With the development of the field of fiber optics and its application to communication systems, a revolution has taken place during the last 35 years or so. Indeed, using glass fibers as the transmission medium and light waves as carrier waves, the terabit barrier of sending information at a rate more than 1 terabit per second (which is roughly equivalent to transmission of about 15 million simultaneous telephone conversations) through one hair-thin optical fiber was crossed during the year 2001. This is certainly one of the extremely important technological achievements of the twentieth century. In this chapter we provide an introduction to propagation through optical fibers and present some of its noncommunication applications. In subsequent chapters we discuss various characteristics of optical fiber that are important for its application to telecommunications and fiber optic sensors.

4.2 GRAHAM BELL'S EXPERIMENT

As discussed in Chapter 3, communication over long distances is achieved using the principle of carrier wave communication. This can be achieved using either analog or digital modulation of the carrier. Since optical beams have frequencies in the range 100 to 1000 THz, the use of such beams as the carrier would imply a tremendously large increase in the information-transmission capacity of the system compared to systems employing radio waves or microwaves; 1 tera hertz (THz) is 1 million megahertz (a medium-wave radio broadcast has a frequency of about 1 MHz). It is the large information-carrying capacity of a light beam that has generated interest among communication engineers to develop a communication system using light waves as carrier waves.

The idea of using light waves for communication can be traced as far back as 1880, when Alexander Graham Bell invented the photophone (Fig. 4.1) shortly after

Fiber Optic Essentials, By K. Thyagarajan and Ajoy Ghatak

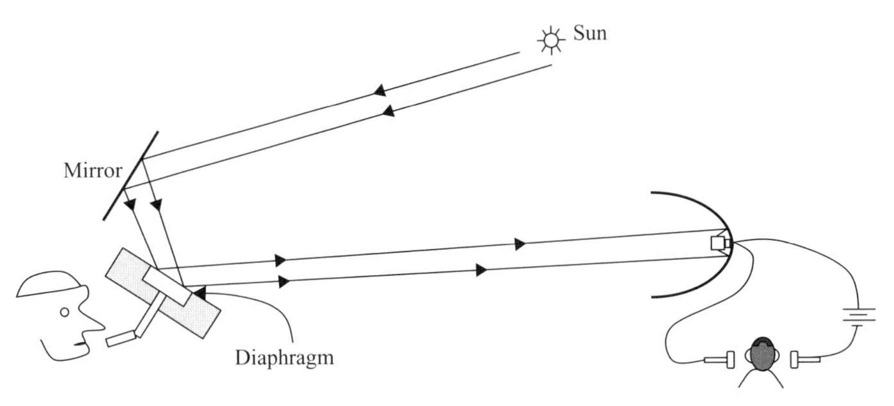

FIGURE 4.1 Schematic of the photophone invented by Alexander Graham Bell. In this system, sunlight was modulated by a diaphragm and transmitted through a distance of about 200 m in air to a receiver containing a selenium cell connected to an earphone.

he invented the telephone[1] in 1876. In this remarkable experiment, speech was transmitted by modulating a light beam, which traveled through air to the receiver. The transmitter consisted of a flexible reflecting diaphragm that could be activated by sound, which was illuminated by sunlight. The reflected light was collimated by a lens, and the reflected beam was received by a parabolic reflector placed at a distance. The parabolic reflector concentrated the light on a photoconducting selenium cell, which forms a part of a circuit with a battery and a receiving earphone. Sound waves present in the vicinity of the diaphragm vibrate the diaphragm, which leads to a variation in the light reflected by the diaphragm. The variation of light falling on the selenium cell changes the electrical conductivity of the cell, which in turn changes the current in the electrical circuit. This changing current reproduces the sound on the earphone. To quote from a book by Maclean (1996): "In 1880 he [Graham Bell] produced his 'photophone' which to the end of his life, he insisted was **'the greatest invention I have ever made, greater than the telephone.** . . .' Unlike the telephone it had no commercial value."

After this beautiful experiment by Alexander Graham Bell communicating using light beams propagating through an open atmosphere, not much work was carried out in the field of optical communications. This was because of the fact that no suitable light source was available that could be modulated to be used as the information carrier. Then in 1960 a revolution occurred: Theodore Maiman fabricated the first laser in 1960. Earlier, no suitable light source was available that could be used reliably as

[1] Actually, according to recent newspaper reports (June 2002), an Italian immigrant, Antonio Meucci, was the inventor of the telephone. According to this report, Antonio Meucci demonstrated his "teletrfono" in New York in 1860. Alexander Graham Bell took out his patent 16 years later. These facts have apparently been recognized by the U.S. Congress.

an information carrier.[2] The advent of lasers thus immediately triggered a great deal of investigation aimed at examining the possibility of building optical analogs of conventional communication systems. The very first such modern optical communication experiments involved laser beam transmission through the atmosphere. However, it was soon realized that laser beams could not be sent in an open atmosphere through reasonably long distances to carry signals, unlike, for example, microwave or radio systems operating at longer wavelengths. This is because a light beam (of wavelength about 1 μm) is severely attenuated and distorted, owing to scattering and absorption by the atmosphere. Thus, for reliable long-distance light-wave communication under terrestrial environments it would be necessary to provide a transmission medium that can protect the signal-carrying light beam from the vagaries of the terrestrial atmosphere.[3]

In 1964, Goubau and Christian at Bell Labs suggested beam waveguides (Fig. 4.2) using lenses for long-distance transmission at optical frequencies. In the same year, Berreman at Bell Labs suggested the use of gas lenses for guiding light beams. In a gas lens a gas in a cylindrical tube is heated along the walls of the tube so that it generates a temperature gradient along the transverse cross section of the tube. Since the refractive index of a gas depends on the temperature, this leads to a variation of the refractive index of the gas in the direction transverse to the axis. This, in turn, leads to focusing effects that result in wave guidance of a light beam. However, both these systems were too bulky to be used in a practical communication system. The appropriate guiding structure is the optical fiber, a hair-thin structure in which the guidance of the light beam through the optical fiber takes place by the phenomenon of total internal reflection (Fig. 4.3). Indeed, it was the important paper of Kao and Hockham in 1966 which suggested that optical fibers based on silica glass could provide the necessary transmission medium if metallic and other impurities could be removed. To quote Kao and Hockham:

> Theoretical and experimental studies indicate that a cladded glass fiber with a core diameter of about one wavelength and an overall diameter of about 1000 wavelengths represents a possible practical optical waveguide with important potential as a new form of communication medium. The refractive index of the core needs to be about 1% higher than that of cladding. However, the attenuation should be around 20 dB/km, which is much higher than the lower limit of loss figure imposed by fundamental mechanisms.

The decibel (dB) scale is defined in Chapter 5; however, we may mention that an attenuation of 20 dB/km implies a power loss by a factor of 100 in traversing through 1 km of the fiber. Now, in 1966, the most transparent glass available at that time had extremely high losses (of more than about 1000 dB/km, implying a power loss by a

[2] We mention here that although incoherent sources such as light-emitting diodes are also often used in present-day optical communication systems, it was the discovery of the laser, which first triggered serious interest in the development of optical communication systems.

[3] Later in this chapter we discuss the emerging area of free-space optics, wherein communication using an open atmosphere rather than optical fiber can be used up to distances of about 4 km. A similar transmission technique via open space is being pursued for applications in space communication between satellites.

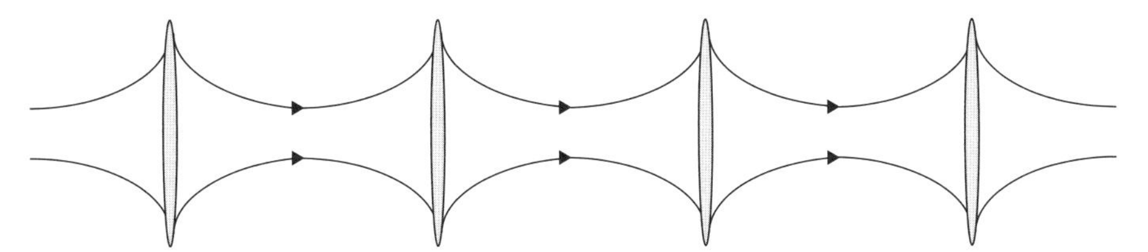

FIGURE 4.2 Beam waveguide consisting of a series of focusing lenses. The lenses compensate for diffraction of the light wave, and thus the wave gets guided.

factor of 100 in traversing through only 20 m of the fiber). The high loss was due primarily to trace amounts of impurities present in the glass. Obviously, this loss is too high even for short distances such as a few hundred meters. The 1966 paper of Kao and Hockham triggered the beginning of serious research in removing traces of impurities present in the glass, which resulted in realization of low-loss optical fibers. In 1970, Kapron, Keck, and Maurer (at Corning Glass) were successful in producing silica fibers with a loss of about 17 dB/km at a He–Ne laser wavelength of 633 nm. Since then, the technology has advanced with tremendous rapidity. By 1985, glass fibers were produced routinely with extremely low losses (<0.25 dB/km, which corresponds to a transmission of 10% of the incident light in 40 km of propagation!);

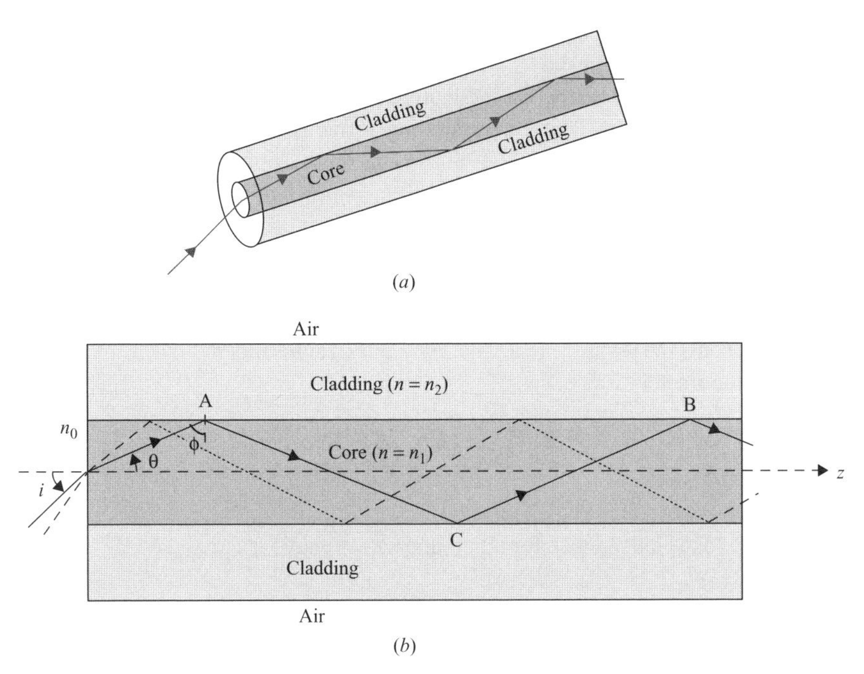

FIGURE 4.3 (*a*) A glass fiber consists of a cylindrical central glass core cladded by a glass of slightly lower refractive index. (*b*) Light rays incident on the core–cladding interface at an angle greater than the critical angle are trapped inside the fiber core.

because of such low losses, the distance between two consecutive repeaters (used for amplifying and reshaping the attenuated signals) could be as large as 250 km.

There was indeed another very important breakthrough in 1970: Alferov in Leningrad and Panish and Hayashi at Bell Labs demonstrated room-temperature operation of semiconductor lasers, which are the sources for optical communication. Indeed, the 2000 Nobel Prize in Physics was awarded to Alferov for his invention. The citation for the prize was:

> In today's society increasing amounts of information flow from our computers out through the optical fibers of the Internet and through our mobile telephones to satellite radio links all over the world. Two simple but fundamental requirements are put on a modern information system for it to be practically useful. It must be fast, so that large volumes of information can be transferred in a short time. The user's apparatus must be small so that there is room for it in offices, homes, briefcases or pockets.
>
> Through their inventions this year's Nobel Laureates in physics have laid a stable foundation for modern information technology.

4.3 OPTICAL FIBER

At the heart of an optical communication system is the optical fiber that acts as the transmission channel carrying a light beam loaded with information. The light beam gets guided through the optical fiber due to the phenomenon of total internal reflection (TIR). Figure 4.3 shows an optical fiber, which consists of a (cylindrical) central dielectric core (of refractive index n_1) cladded by a material of slightly lower refractive index n_2 ($<n_1$). The corresponding refractive index distribution (in the transverse direction) is given by

$$n = \begin{cases} n_1 & r < a \\ n_2 & r > a \end{cases} \tag{4.1}$$

where n_1 and n_2 ($<n_1$) represent the refractive indices of core and cladding, respectively, r represents the distance from the axis, and a represents the radius of the core. We define a parameter Δ as

$$\Delta \equiv \frac{n_1^2 - n_2^2}{2n_1^2} \tag{4.2}$$

When $\Delta \ll 1$ (as is indeed true for silica fibers), we may write

$$\Delta \approx \frac{n_1 - n_2}{n_2} \approx \frac{n_1 - n_2}{n_1} \tag{4.3}$$

The necessity for a cladded fiber (Fig. 4.3) rather than a bare fiber (i.e., without cladding), arises from the fact that for transmission of light from one place to another, the fiber must be supported, and supporting structures may distort the fiber considerably, thereby affecting the guidance of the light wave. This can be avoided by choosing a sufficiently thick cladding. Further, in a fiber bundle, in the absence of cladding, light can leak through from one fiber to another, leading to possible crosstalk among the information carried between two different fibers. The idea of adding a second layer of glass (i.e., the cladding) came in 1955 independently from Hopkins and Kapany in the United Kingdom and van Heel in Holland. However, during that time the use of optical fibers was primarily in image transmission rather than in communications. Indeed, the early pioneering works in fiber optics (in the 1950s) by Hopkins, Kapany, and van Heel led to the use of fiber in optical devices.

Now, for a ray entering the fiber, if the angle of incidence ϕ at the core–cladding interface (see Fig. 4.3*b*) is greater than the critical angle ϕ_c [$= \sin^{-1}(n_2/n_1)$], the ray will undergo TIR at that interface; (we discussed TIR briefly in Chapter 2). We mention here that the concept of rays is really valid only for multimode fibers where the core radius a is large (≈ 25 μm or more); we give a more detailed description of single- and multimode fibers in the next section. For a typical (multimoded) fiber, $a \approx 25$ μm, $n_2 \approx 1.45$ (pure silica), and $\Delta \approx 0.01$, giving a core index of $n_1 \approx 1.465$. The cladding is usually pure silica, while the core is usually silica doped with germanium; doping by germanium results in an increase in the refractive index. Now, because of the cylindrical symmetry in the fiber structure, this ray will also suffer TIR at the lower interface and therefore get guided through the core by repeated total internal reflections. Figure 4.4 shows light propagating through an optical fiber. The fiber is visible due to the phenomenon of Rayleigh scattering, which scatters a tiny part of the light propagating through the fiber and makes the fiber visible. Rayleigh scattering is the phenomenon responsible for the blue color of the sky and the red color of the rising or setting sun. Even for a bent fiber, light guidance can occur through multiple total internal reflections, as can be seen from Fig. 4.4.

FIGURE 4.4 Long, thin optical fiber carrying a light beam. The fiber is visible due to Rayleigh-scattered light.

4.4 WHY GLASS FIBERS?

Why are optical fibers made of glass? Quoting W. A. Gambling (1986), a pioneer in the field of fiber optics, we note that glass is a remarkable material that has been in use in "pure" form for at least 9000 years. The compositions remained relatively unchanged for millennia, and its uses have been widespread. The three most important properties of glass, which make it of unprecedented value, are:

1. Through a wide range of accessible temperatures in which the viscosity of glass is variable and can be well controlled (unlike most materials, such as water and metals, which remain liquid until they are cooled down to their freezing temperatures and then suddenly become solid), glass does not solidify at a discrete freezing temperature but gradually becomes stiffer and stiffer and eventually becomes hard. In the transition region it can easily be drawn into a thin fiber.
2. Highly pure silica is characterized by extremely low loss (i.e., it is highly transparent). Today, in most commercially available silica fibers, 96% of the power gets transmitted after propagating through 1 km of optical fiber. This indeed represents a truly remarkable achievement.
3. The intrinsic strength of glass is about 2,000,000 lb/in^2, so that a glass fiber of the type used in the telephone network, which has a diameter (125 μm) twice the thickness of a human hair, can support a load of 40 lb.

Although for a common person, glass looks fragile, glass fibers are indeed extremely strong. It is the exposure of the glass to the external atmosphere that leads to the formation of cracks, which then results in fracture. In the case of optical fibers, these are drawn in an extremely clean environment and are coated with polymers as they are being drawn. Covering the glass fiber with polymer does not permit contact with the atmosphere and gives it the protection.

4.5 FIBER OPTIC BUNDLE

A large number of optical fibers assembled together form a *bundle*. If the fibers are not aligned (i.e., they are all jumbled up), the bundle is said to form an *incoherent bundle*. Such bundles can be used to transmit light from one point to another along a flexible path into regions where normal access is not available. Incoherent bundles are also used in illumination, such as in traffic lights, road signs, and buildings where the light source must be separated because of the relative inaccessibility and wherein fibers used to guide light from a lamp. They can also be used as cold light sources (i.e., light sources giving light but no heat) by cutting off heat radiation using a filter at the input to the fiber bundle (Fig. 4.5). The light emerging from the bundle is also free from ultraviolet radiation and is suitable for illumination of exhibits in museums. Another very interesting application is the use of such bundles to bring sunlight into working areas, which can give light without both heat and harmful ultraviolet

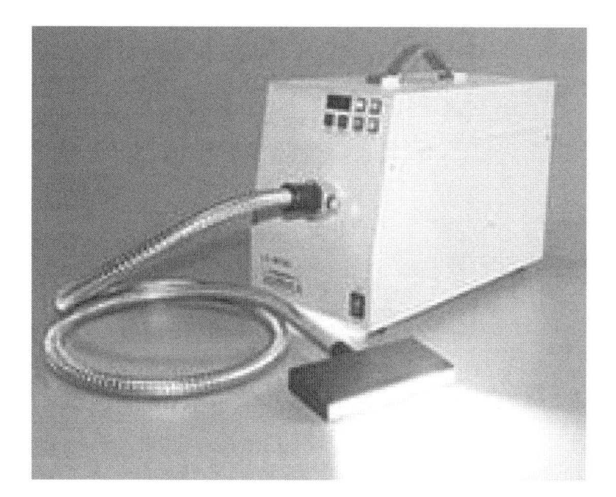

FIGURE 4.5 Cold light source. (Ref: Sumita Optical Glass, Inc.)

radiations. There are companies that offer such a source. Decorative applications of fiber optic illuminators are also very interesting. Such illuminators are attractive since they provide safety from electrical hazards and project less load on air conditioning, due to the removal of heat components from light. For example, special effects can be incorporated for color changes for specific applications.

Fiber optic signage is another interesting application wherein instead of using neon bulbs, fibers can be used to create imaginative designs. Typical applications include cove lighting, walkway lighting, traffic signals (Fig. 4.6), and entertainment

FIGURE 4.6 Fiber optic traffic sign. (From http://www.nationalssc.com/.)

FIGURE 4.7 Fiber optic display of the 1996 Olympics showing the Olympic torch and flame. (From http://www.lightdesignsystems.com/fiber.htm.)

illumination. In fact, for the 1996 Olympics, a fiber optic display showing the Olympic torch and flame was used to great effect (Fig. 4.7). Indeed, the world's largest fiber optic lighting, in Times Square in New York City, has about 65,000 points of light (Ref: http://www.fibercreations.com/index.html).

There is also interest in fluorescent or scintillating fibers, which are ordinary plastic fibers doped with special elements. When light falls on a *fluorescent fiber*, it excites the dopants to an excited state, and when the atoms relax, they emit light. Part of this light is collected within the fiber core and is transmitted to both ends of the fiber. In *scintillating fibers*, light gets generated when radiations such as α or β particles or γ rays are incident on the fiber. Many applications are envisaged for fluorescent fibers, such as in intrusion alarms and for size determination, due particularly to their ability to capture light that falls anywhere along the length of the fiber. Scintillating optical fibers are used in particle energy measurement and track detection.

If the optical fibers are aligned properly (i.e., if the relative positions of the fibers in the input and output ends are the same), they are said to form a *coherent bundle*. Now, if a particular fiber is illuminated at one of its ends, there will be a bright spot at the other end of the same fiber at the same position; thus, a coherent bundle can transmit an image from one end to another (Fig. 4.8). On the other hand, in an incoherent bundle the output image will be scrambled. Because of this property, an incoherent bundle can be used as a coder, and the image transmitted can be decoded by using a similar bundle at the output end. In a bundle, since there can be hundreds of thousands of fibers, decoding without the original bundle configuration should be extremely difficult.

Perhaps the most important application of a coherent bundle is in a fiber optic endoscope, where it can be put inside a human body and the interior of the body can

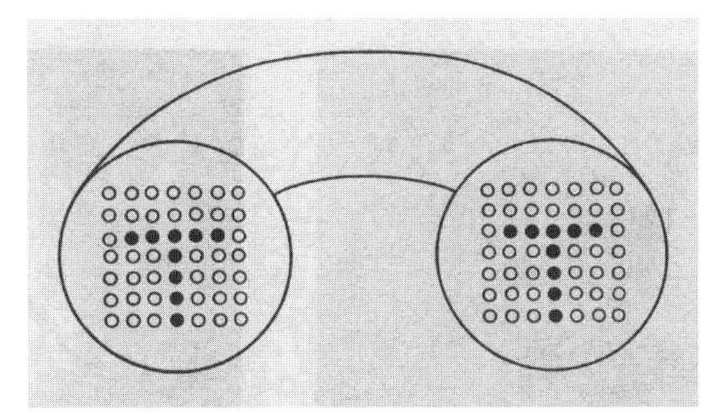

FIGURE 4.8 Bundle of aligned fibers. A bright (or dark) spot at the input end of the fiber produces a bright (or dark) spot at the output end. Thus, an image will be transmitted (in the form of bright and dark spots) through a bundle of aligned fibers.

be viewed from outside; for illuminating the portion that is to be seen, the bundle is enclosed in a sheath of fibers that carry light from outside to the interior of the body (Fig. 4.9). A state-of-the-art fiberscope can have about 10,000 fibers, which would form a bundle about 1 mm in diameter capable of resolving objects 70 μm across.

It is of interest to mention here that the retina of the human eye consists of a large number of rods and cones, which have the same type of structure as the optical fiber: They consist of dielectric cylindrical rods surrounded by another dielectric of slightly lower refractive index (Fig. 4.10). The core diameters are in the range of a few micro meters. The light absorbed in these *light guides* generates electrical signals which are transmitted to the brain through various nerves, which enables us to see the world outside. Recent discoveries have shown that spicules of the deep-sea "glass" sponge *Euplectella* have a remarkable similarity to manufactured optical fibers. Figure 4.11*a* shows the cross section of a spicule with various regions marked,

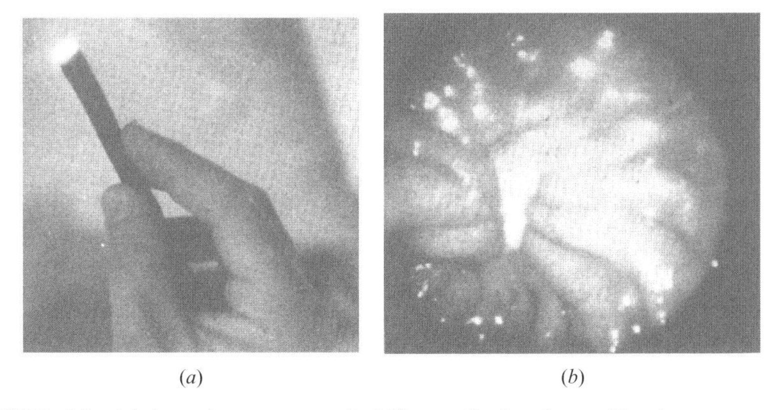

(*a*) (*b*)

FIGURE 4.9 (*a*) An endoscope, an optical fiber medical probe, enables doctors to examine the inner parts of the human body. (*b*) A stomach ulcer as seen through an endoscope. (Courtesy of the United States Information Service, New Delhi, India.)

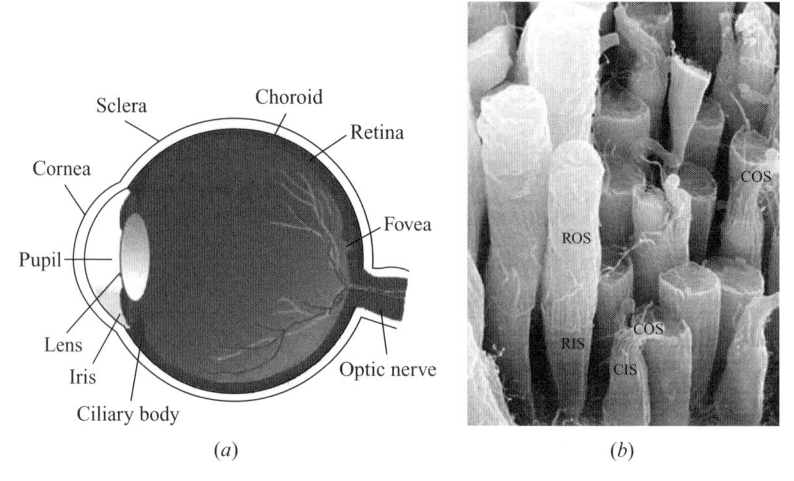

FIGURE 4.10 The rods and cones of the eye behave like optical waveguides. [(*a*) Adapted from http://webvision.med.utah.edu/imageswv/sagitta2.jpeg; (*b*) adapted from http://www.uchc.edu/dsp/image/frog_retina.jpg.]

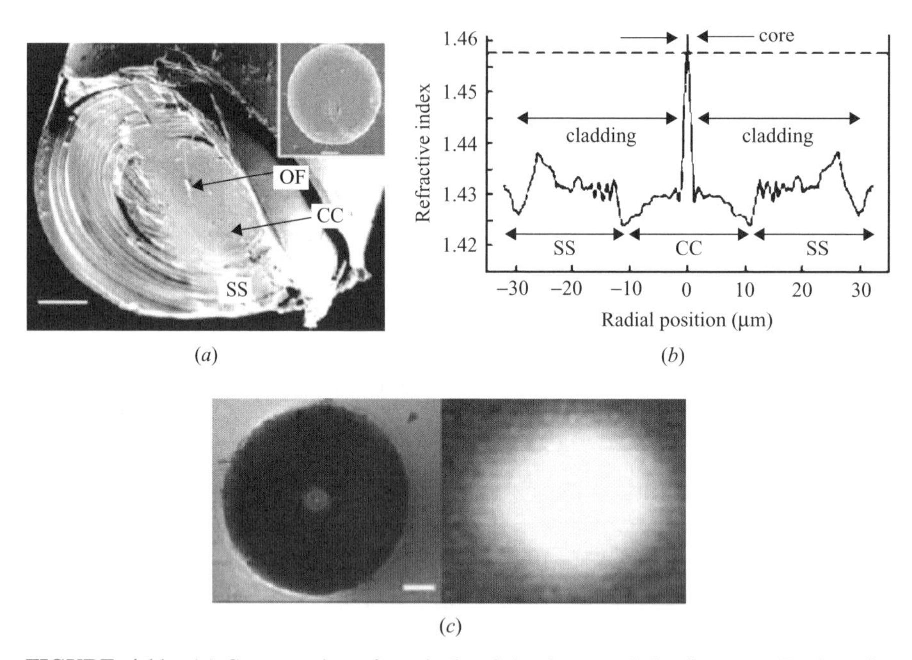

FIGURE 4.11 (*a*) Cross section of a spicule of the deep-sea "glass" sponge *Euplectella*, also referred to as the Venus flower basket. OF, organic filaments; SS, outer striated shell; CC, central cylinder. (*b*) Corresponding transverse refractive index profile showing a core and a cladding. The dashed line indicates the refractive index of vitreous silica. (*c*) Light emerging from the fiber. (Adapted from Sundar et al., 2003.)

and Fig. 4.11*b* shows the transverse refractive index profile of the spicule showing clearly a waveguide structure capable of guiding light. These structures act like single- and multimode waveguides (Fig. 4.11*c*). The interesting aspects of these naturally occurring optical fibers is that they are being manufactured in ambient conditions, whereas manufactured optical fibers require very high temperatures (about 1700°C). As discussed later in the chapter, organic ligands on the exterior of the fiber seem to protect it and provide an effective crack-arresting mechanism, and the fibers seem to be doped with specialized impurities such as sodium that improve the refractive index profile and hence the waveguiding properties. Studies of such fibers could lead to better optical fibers for telecommunication applications.

4.6 NUMERICAL APERTURE OF THE FIBER

We return to Fig. 4.3*b* and consider a ray that is incident on the entrance aperture of the fiber, making an angle i with the axis. Let the refracted ray make an angle θ with the axis. Assuming the outside medium to have a refractive index n_0 (which for most practical cases is unity), we get

$$\frac{\sin i}{\sin \theta} = \frac{n_1}{n_0} \tag{4.4}$$

Obviously, if this ray has to suffer total internal reflection at the core–cladding interface,

$$\sin \phi (= \cos \theta) > \frac{n_2}{n_1} \tag{4.5}$$

Thus,

$$\sin \theta < \left[1 - \left(\frac{n_2}{n_1} \right)^2 \right]^{\frac{1}{2}} \tag{4.6}$$

and we must have $i < i_m$, where

$$\sin i_m = \left(n_1^2 - n_2^2 \right)^{\frac{1}{2}} = n_1 \sqrt{2\Delta} \tag{4.7}$$

and we have assumed that $n_0 = 1$ (i.e., the outside medium is assumed to be air). Thus, if a cone of light is incident on one end of the fiber, it will be guided through it provided that the semiangle of the cone is less than i_m. This angle is a measure of the light-gathering power of the fiber, and as such, one defines the *numerical aperture* (NA) of the fiber by the equation

$$\text{NA} = \sin i_m = \sqrt{n_1^2 - n_2^2} \approx n_1 \sqrt{2\Delta} \tag{4.8}$$

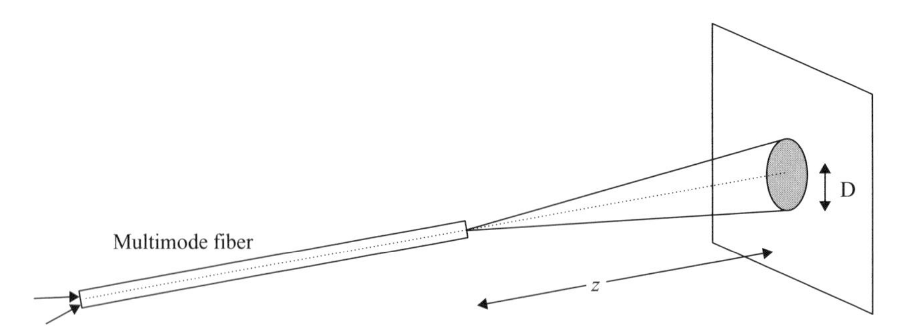

FIGURE 4.12 Diameter D of the spot on a screen placed a distance z from the output end of a multimode fiber can be used to estimate the NA of the fiber.

Example 4.1 For a typical step-index (multimode) fiber with $n_1 \approx 1.45$ and $\Delta \approx 0.01$, we get

$$\sin i_m \approx 0.205 \Rightarrow i_m \approx 12^\circ$$

Now, in a short length of multimode optical fiber, if all rays between $i = 0$ and i_m are launched, the light coming out of the fiber will also appear as a cone of semiangle i_m emanating from the fiber end. If we now allow this beam to fall normally on white paper and measure its diameter, we can easily calculate the NA of the fiber. Several concentric circles of increasing radii—say, starting from 0.5 to 1.5 cm—are drawn on a small paper screen and the screen is positioned in the far field such that at the output end, the axis of the fiber passes perpendicularly through the center of these circles on the screen (Fig. 4.12). The fiber end, which is mounted on an XYZ-stack, is moved slightly toward or away from the screen so that one of the circles just circumscribes the far-field radiation spot. The distance z between the fiber end and the screen and the diameter D of the coinciding circle are measured accurately. The NA is calculated using the equation

$$\mathrm{NA} = \sin i_m = \sin\left(\tan^{-1}\frac{D}{2z}\right) \tag{4.9}$$

4.7 MULTIMODE AND SINGLE-MODE FIBERS

In this section we examine the difference between multimode and single-mode fibers, both of which are used in optical communication systems. First, we introduce the concept of modes: A *mode* is a transverse field distribution that propagates along the fiber without any change in its field distribution except for a change in phase. Mathematically, it is defined by

$$\Psi(x,y,z,t) = \psi(x,y)e^{i(\omega t - \beta z)} \tag{4.10}$$

where $\psi(x,y)$ represents the transverse field profile and β represents the propagation constant; propagation of the mode is in the z direction. The quantity β is similar to k for a plane wave given by $e^{i(\omega t-kz)}$. We note that the propagation constant k of a plane wave in a medium of refractive index n is $(\omega/c)\,n$, where ω is the frequency of the electromagnetic wave. In a similar fashion we can define an *effective index* of a mode having a propagation constant β by the equation

$$n_{\mathrm{eff}} = \frac{\beta}{\omega/c} \tag{4.11}$$

A mode with a propagation constant β propagates as if it were a plane wave propagating in a medium of refractive index n_{eff}: hence the name *effective index*. The effective index of all modes guided by an optical fiber lies between the core and cladding refractive indices. As we will see in Chapter 7, the wavelength variation of the effective index determines the pulse broadening in single-mode optical fibers. Thus, the effective index is a very important quantity for a single-mode fiber.

A waveguide such as an optical fiber is characterized by a finite number of modes which are guided by the waveguide; each mode is described by a definite transverse field distribution $\psi(x,y)$, corresponding to a definite value of β. The precise form of $\psi(x,y)$ (and the corresponding value of the propagation constant β) is obtained by solving Maxwell's equations. This topic is beyond the scope of the book, but interested readers will find a number of excellent textbooks listed at the end of the book.

Figure 4.13 shows transverse field patterns corresponding to some modes of an optical fiber. The quantities l and m represent numbers that specify the mode, which

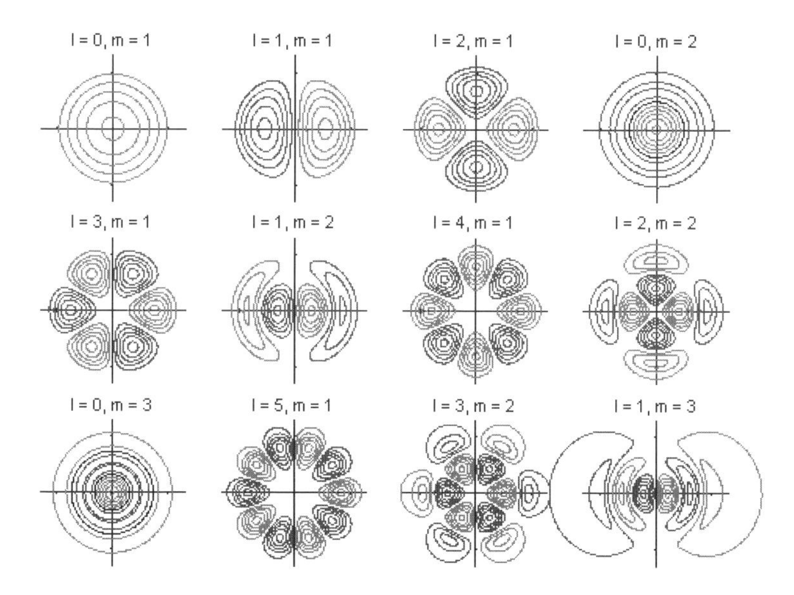

FIGURE 4.13 Field patterns of some low-order guided modes. (After http://www.rp-photonics.com/fibers.html, Rutger Paschotta.)

are called *linearly polarized* (LP_{lm}) *modes*;. These resemble the vibrational pattern of a drum. Each mode shown is characterized by a different propagation constant β. For a step-index fiber, defined by Eq. (4.1), we define a dimensionless parameter V by the equation

$$V = \frac{2\pi}{\lambda_0} a \sqrt{n_1^2 - n_2^2} = \frac{2\pi}{\lambda_0} a n_1 \sqrt{2\Delta} \tag{4.12}$$

λ_0 is the free-space wavelength of the light beam and Δ is defined by Eq. (4.2). The parameter V (which also depends on the operating wavelength λ_0), known as the *waveguide parameter*, is an extremely important quantity characterizing an optical fiber. For a step-index fiber [see Eq. (4.1)], if

$$V < 2.4045 \tag{4.13a}$$

the fiber is said to be a *single-mode fiber*. Such a fiber supports only the LP_{oi} mode with a field distribution shown in the top left hand corner of Fig. 4.13. For a given fiber, the wavelength for which $V = 2.4045$ is known as the *cutoff wavelength* and is denoted by λ_c, and the fiber will be single-moded for $\lambda_0 > \lambda_c$ (see Example 4.2). However, if

$$V \geq 10 \tag{4.13b}$$

the fiber is said to be a highly *multimoded fiber*.

Earlier, when discussing step-index fibers, we considered light propagation inside the fiber as a set of many rays bouncing back and forth at the core–cladding interface (Fig. 4.3*b*), and the angle θ could take *all* possible values from 0 (corresponding to a ray propagating parallel to the z axis) to $\cos^{-1}(n_2/n_1)$ (corresponding to a ray incident at the critical angle on the core–cladding interface). However, we can demonstrate by experiment that only discrete values of θ are allowed. Figure 4.14 shows an arrangement that demonstrates this feature. A prism of high refractive index is placed very close to the core of an optical waveguide. The waveguide could correspond to an optical fiber in which most of the cladding is removed from one side so that the core is accessible. This can be achieved either by polishing the fiber from its side or by using etchants that etch the cladding. Now when the prism is placed very close to the core and light is coupled into the core from one end, we find that the light emerges from the prism only at discrete angles, each angle corresponding to a particular mode of the optical fiber. Each discrete ray path corresponds to the mode of the waveguide. Indeed, in the experiment shown in Fig. 4.14, if we measure the discrete values of θ, we can obtain the discrete values of the propagation constant β.

Now, if $V < 2.4045$, we will have only one discrete ray path and the fiber is a single-mode fiber. For such a fiber, one should use wave optics to study the propagation characteristics of the optical fiber, and ray optics will not be applicable. Further, for a single-mode fiber there is only the guided mode, and the transverse field distribution $\psi(x,y)$ associated with the mode is approximately Gaussian; we discuss this

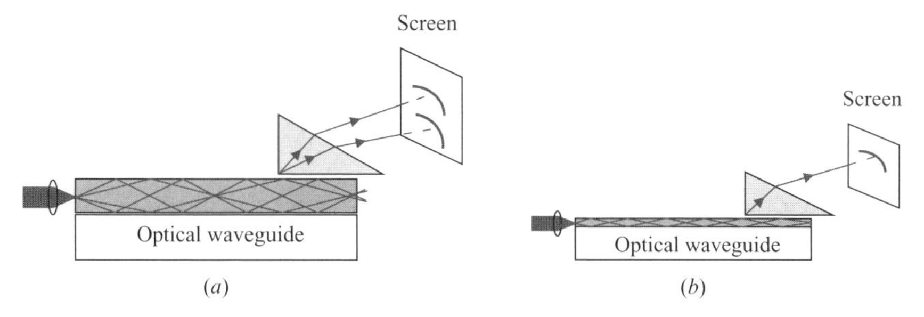

FIGURE 4.14 A laser beam is focused at the entrance aperture of a waveguide. (*a*) Each discrete ray path corresponds to a mode of the waveguide. (*b*) For a single-mode waveguide there is only one discrete value of θ.

in the next section. On the other hand, for $V \geq 10$, the number of modes is given approximately by $\frac{1}{2}V^2$ and the fiber is a multimode fiber. In a multimode fiber, different modes travel with different group velocities, leading to what is known as *intermodal dispersion*; in the language of ray optics, this is known as *ray dispersion* because different rays take different amounts of time in propagating through the fiber (see Chapter 6). Indeed, in a highly multimoded fiber of many modes we can use ray optics to calculate pulse dispersion.

Example 4.2 Consider a step-index fiber with $n_2 = 1.447$, $\Delta = 0.003$, and $a = 4.2$ μm. Thus,

$$V = \frac{2\pi}{\lambda_0}(4.2)(1.447)\left(\sqrt{0.006}\right) \approx \frac{2.958}{\lambda_0}$$

where λ_0 is measured in micrometer. Thus, for

$$\lambda_0 > \frac{2.958}{2.4045} \approx 1.23\ \mu\text{m}$$

the fiber will support only one mode. Thus, in this example, the *cutoff wavelength* $\lambda_c = 1.23$ μm and the fiber will be single moded for $\lambda_0 > 1.23$ μm. If the fiber is operating at 1300 nm

$$V = \frac{2.958}{1.3} \approx 2.275$$

and the fiber will support a single mode.

Example 4.3 For reasons that we discuss later, the fibers used in fourth generation optical communication systems (operating at 1.55 μm) have a small core radius value and a large Δ value. A typical fiber would have $n_2 = 1.444$, $\Delta = 0.0075$, and $a =$ 2.3 μm, for which

$$V = \frac{2\pi}{\lambda_0}(2.3)(1.444)\sqrt{0.015} \approx \frac{2.556}{\lambda_0}$$

and therefore the cutoff wavelength will be $\lambda_c = 2.556/2.4045 = 1.06$ μm. If we operate at $\lambda_0 = 1.55$ μm,

$$V \approx 1.649$$

and the fiber will support a single mode.

4.8 STEP- AND GRADED-INDEX MULTIMODE FIBERS

Step-index multimode fibers have a core made of uniform refractive index medium. This leads to a step discontinuity in the refractive index profile, and hence such fibers are referred to as step-index fibers. The transverse cross section of a typical step-index multimode fiber is shown in Fig. 4.15*a*. The cladding is usually pure silica and the core refractive index is about 1% more than the cladding refractive index. The

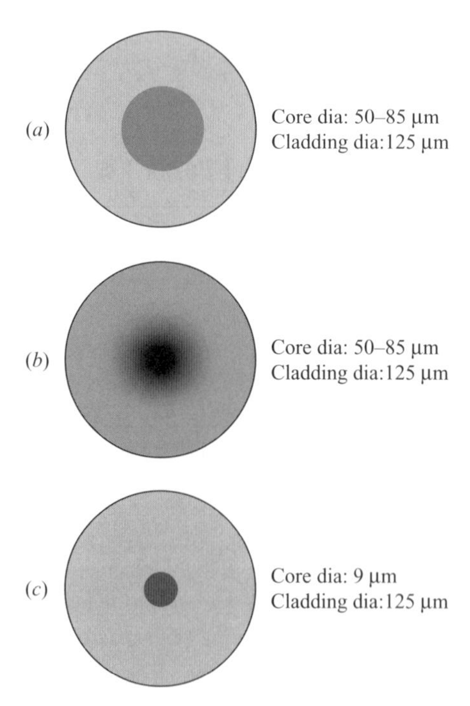

FIGURE 4.15 Transverse cross sections and typical dimensions of (*a*) step-index multimode, (*b*) graded-index multimode, and (*c*) single-mode fibers.

propagation of light through such fibers can be described in terms of different rays, which undergo total internal reflection at the core–cladding interface.

Graded-index multimode fibers are characterized by cores in which the refractive index decreases as we move away from the fiber axis. The transverse refractive index variation is usually very nearly parabolic; that is, the refractive index decreases quadratically as we move away from the axis of the fiber (Fig. 4.15*b*)

$$n^2(r) = \begin{cases} n_1^2\left[1 - 2\Delta\left(\frac{r}{a}\right)^2\right] & 0 < r < a \quad \text{core} \\ n_2^2 = n_1^2(1 - 2\Delta) & r > a \quad \text{cladding} \end{cases} \tag{4.14}$$

with Δ again as defined in Eq. (4.2). For a typical (multimode) parabolic index silica fiber, $\Delta \approx 0.01$, $n_2 \approx 1.45$, and $a \approx 25\ \mu\text{m}$. Due to the graded refractive index profile within the fiber core, light propagates through such fibers along sinusoidal ray paths, as shown in Fig. 4.16*b*. This type of variation leads to increased information-carrying capacity (see Chapter 6). Figure 4.17 shows the measured refractive index profile of a typical graded-index optical fiber.

Due to the large core diameters and relatively large refractive index difference between core and cladding, one can use inexpensive light-emitting diodes (LEDs) rather than the more expensive laser diodes (LDs). Thus, multimode fibers are the fibers of choice in applications involving light gathering or illumination. Multimode fibers can also be used to set up communication over short distances, and for such applications,

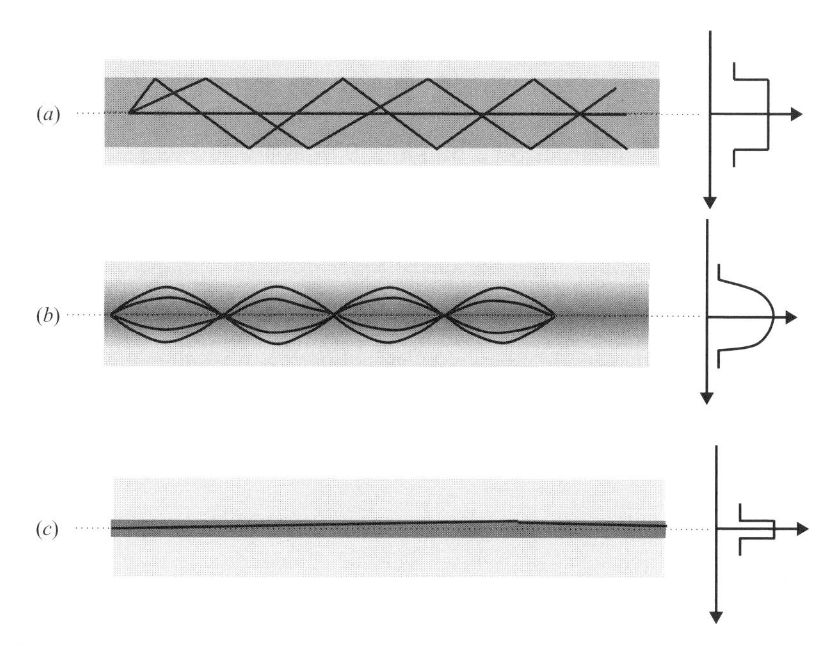

FIGURE 4.16 Ray paths in (*a*) step-index multimode, (*b*) graded-index multimode, and (*c*) single-mode optical fibers.

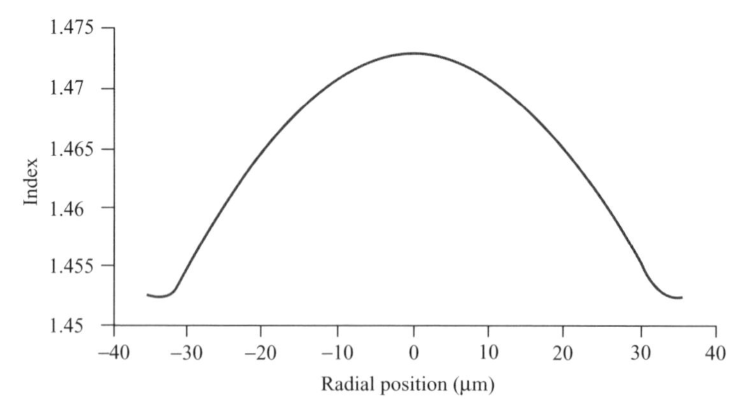

FIGURE 4.17 Typical refractive index profile of a graded-index multimode fiber optimized for use with laser sources. (Adapted from Pondillo, 2001.)

they can be less expensive, due to greater tolerance in various components such as LEDs and connectors. Single-mode fibers are the fibers of choice in long-distance communication and in fiber optic sensors. Due to the small core diameter and smaller acceptance angle, single-mode fiber applications use laser diodes as sources.

Compared to multimode fibers, in a single-mode fiber the core radius is approximately 5 μm, $n_2 \approx 1.45$ (pure silica), and $n_1 \approx 1.456$; thus, $\Delta \approx 0.004$ and the core and cladding indices differ by about 0.4%. In both multimode and single-mode fibers the outer diameter of the cladding is 125 μm. For such core dimensions, light propagation cannot be described using simple ray optics, and the more rigorous wave theory has to be used.

Spot Size of the Fundamental Mode in a Single-Mode Fiber

As mentioned earlier, a single-mode fiber supports only one mode that propagates through the fiber, referred to as the *fundamental mode* of the fiber. The transverse field distribution associated with the fundamental mode of a single-mode fiber is an extremely important quantity that determines various important parameters, such as splice loss at joints, launching efficiencies, and bending loss. For most single-mode fibers, the fundamental-mode field distributions can be well approximated by a Gaussian function, which may be written in the form

$$\psi(x, y) = Ae^{-(x^2+y^2)/w^2} = Ae^{-r^2/w^2} \tag{4.15}$$

where w is referred to as the *spot size* of the mode field pattern and 2w is defined as the mode field diameter (MFD). For a step-index single-mode fiber, one has the following empirical expression for w:

$$\frac{w}{a} \approx 0.65 + \frac{1.619}{V^{3/2}} + \frac{2.879}{V^6} \qquad 0.8 \le V \le 2.5 \tag{4.16}$$

where a is the core radius. We may mention here that the light coming from a He–Ne laser (or a laser pointer) has a transverse intensity distribution very similar to that coming from a single-mode fiber except that the spot size is much larger. The standard single-mode fiber, designated as G.652 fiber for operation at 1310 nm, has an MFD of 9.2 ± 0.4 μm and an MFD of 10.4 ± 0.8 μm at 1550 nm.

Example 4.4 Consider a step-index fiber operating at 1300 nm with $n_2 = 1.447$, $\Delta = 0.003$, and $a = 4.2$ μm (see Example 4.2). Thus, $V \approx 2.28$, giving $w \approx 4.8$ μm. The same fiber will have a V value of 1.908 at $\lambda_0 = 1550$ nm, giving a value of the spot size ≈ 5.5 μm. *Thus, in general, the spot size increases with wavelength.*

Example 4.5 Consider a step-index fiber operating at 1550 nm with $n_2 = 1.444$, $\Delta = 0.0075$, and $a = 2.3$ μm (see Example 4.3). Thus, $V \approx 1.65$, giving $w \approx 3.6$ μm. The same fiber will have a V value of 1.97 at $\lambda_0 = 1300$ nm, giving a spot size value ≈ 3.0 μm.

Example 4.6 Consider two single-mode fibers, one with $a = 5$ μm and NA = 0.1 and the other with $a = 2.5$ μm and NA = 0.2. Both fibers have a V value of 2.417 at 1300 nm. However, their spot sizes are 5.48 and 2.74 μm, respectively. Thus, higher, NA fibers have a smaller spot size for the same V number. A smaller spot size implies tighter mode confinement and results in lower bend-induced loss.

4.9 SPLICE LOSS DUE TO TRANSVERSE MISALIGNMENT IN A SINGLE-MODE FIBER

The most common misalignment at a joint between similar fibers is transverse misalignment, similar to that shown in Fig. 4.18. Corresponding to transverse misalignment of u, the loss in decibels is given by

$$\alpha(\text{dB}) \approx 4.34 \left(\frac{u}{w}\right)^2 \tag{4.17}$$

Thus, a larger value of w will lead to a greater tolerance for transverse misalignment. For $w \approx 5$ μm and a transverse offset of 1 μm, loss at the joint will be approximately 0.18 dB; on the other hand, for $w \approx 3$ μm, a transverse offset of 1 μm will result in a loss of about 0.5 dB.

Example 4.7 For a single-mode fiber operating at 1300 nm, $w = 5$ μm, and if the splice loss is to be below 0.1 dB, then from Eq. (4.17) we obtain $u < 0.76$ μm.

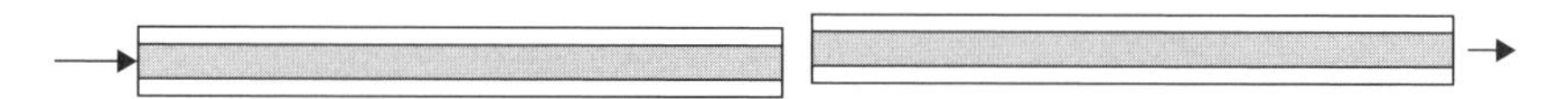

FIGURE 4.18 Transverse misalignment of two identical single-mode fibers.

Thus, for a low-loss joint, the transverse alignment is very critical, and connectors for single-mode fibers require precision matching and positioning to achieve low loss.

4.10 FABRICATION OF OPTICAL FIBERS

As discussed earlier, optical fibers used in communication are made of silica, which is simply silicon dioxide, a major component of sand. Optical fibers must be constructed from ultra pure starting materials so that the resulting optical fiber is almost totally free of impurities. Also, the fiber geometry needs to be maintained along multiple-kilometer lengths of the fiber. The transverse refractive index profile of the fiber determines its information-carrying capacity and needs to be controlled very carefully. To achieve these characteristics, different methods of optical fiber manufacture have been developed. Here we discuss briefly one of the most important methods of manufacture: the modified chemical vapor deposition (MCVD) method developed by Bell Labs in 1974, which has two major stages.

Preform Fabrication

The process begins with fabrication of the *preform*, which is a cylinder about 10 to 20 cm in diameter and about 1 m long. This preform has the same refractive index profile as the optical fiber that is to be fabricated, but on a much larger scale. The refractive index profile, geometry of the core and cladding, core concentricity, and other parameters desired are all decided by the preform. Hence, preform fabrication is a very important stage and the propagation characteristics of the final fiber are determined primarily by the quality of the preform.

Figure 4.19 shows a setup to fabricate optical fiber preform using the MCVD method. An ultra pure fused silica tube is loaded on a glass-working lathe through

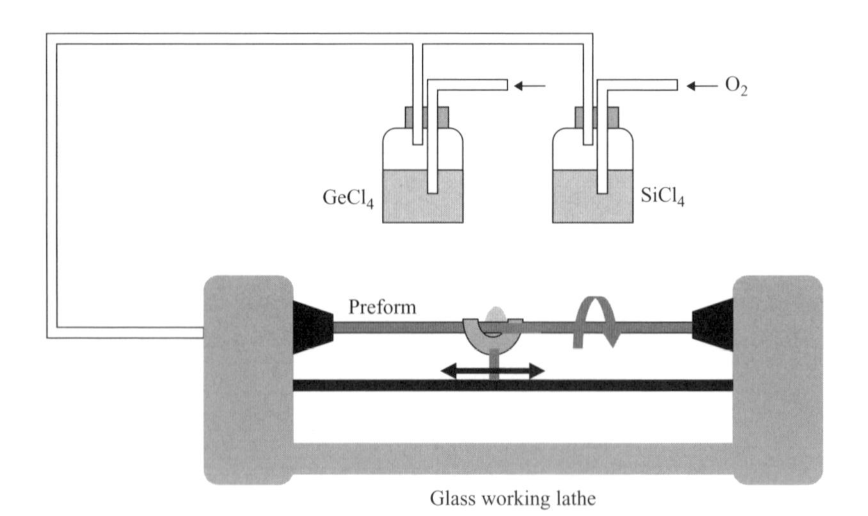

FIGURE 4.19 Preform fabrication in a glass-working lathe.

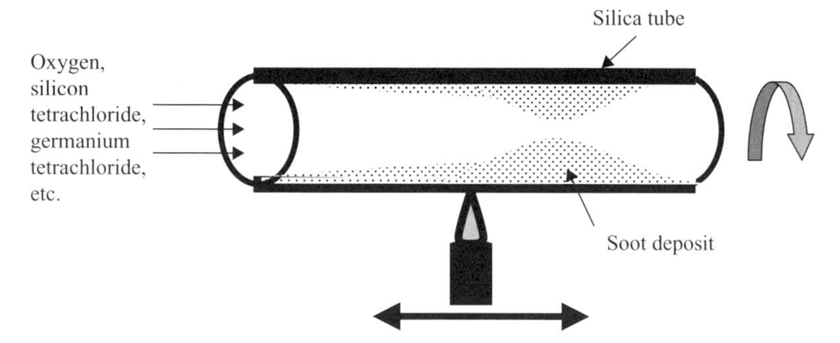

FIGURE 4.20 Soot is formed during the reaction stage that takes place at the high-temperature point created by the burner.

which ultrapure reactant gases of silicon tetrachloride, oxygen, germanium tetrachloride, and so on, can be sent. The fused silica tube is rotated along its axis and a high-temperature (about 1600°C) burner heats the tube within a narrow zone while traveling back and forth along its length. When the reactant gases pass through the heated zone of the silica tube, oxidation takes place. Thus, if the gases are silicon tetrachloride and oxygen (obtained by bubbling oxygen through silicon tetrachloride solution), the oxidation reaction produces silicon dioxide in the form of fine soot (about 0.1-μm-sized particles). As the silica tube rotates, the soot gets deposited uniformly all along the inner sidewalls of the silica tube (Fig. 4.20). As the burner passes back and forth, multiple layers of soot can be deposited. Deposition of pure silica in this fashion will finally form the cladding of the fiber. To form the core, other gases, such as germanium tetrachloride, are mixed with silicon tetrachloride and oxygen in appropriate proportions so that the soot formed contains germanium dioxide along with silicon dioxide. Thus, these layers would consist of silica doped with germanium. As the doping results in an increase in the refractive index, these layers would correspond to the core of the fiber. It is at this stage that the concentration of germanium (or any other element) can be controlled very precisely layer by layer to achieve the desired refractive index profile of the final fiber. Figure 4.21 shows an actual preform fabrication unit.

After depositing the required number of layers, the preform is sintered: The temperature of the burner is increased to about 2000°C and the preform is moved slowly, resulting in collapse of the tube into a rod. At the same time, the soot solidifies and forms glass.

Fiber Fabrication

Once the preform has been fabricated, measurements are made to determine its transverse refractive index profile, uniformity along the length, dimensions, and so on. From the measurements it is possible to estimate the final propagation characteristics of the optical fiber that would be drawn from the preform. Depending on the performance estimated, either the preform is rejected (if it does not meet the requirements) or it is drawn into the optical fiber.

FIGURE 4.21 Preform fabrication setup. (Courtesy of Dr. S. K. Bhadra, CGCRI, Kolkata, India.)

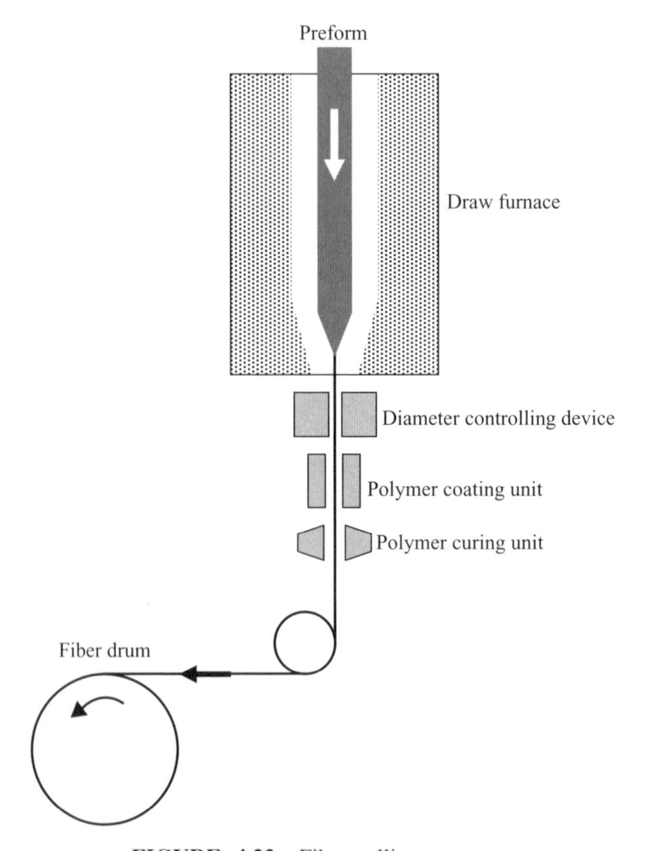

FIGURE 4.22 Fiber pulling apparatus.

Figure 4.22 shows a fiber drawing unit. The preform is lowered into the drawing furnace so that the tip of the preform melts and results in flow of the glass in the form of a fine fiber. The fiber is rolled into a drum and pulled in a uniform fashion to maintain the overall dimensions of the fiber. Before reaching the drum, online instruments measure the diameter of the fiber. The measurement unit has feedback on the pulling speed of the fiber so that the diameter can be controlled very precisely. Thus, if the diameter is slightly larger, the pulling speed is increased, and if the diameter is slightly smaller, the pulling speed is reduced to maintain the correct set diameter. The fiber is also coated online with a suitable polymer, which is cured by ultraviolet lamps or by heating. This very important step is required to protect the freshly drawn fiber from exposure to the environment, which can degrade the strength of the fiber to a very large extent. It also gives extra strength to the fiber, resistance to abrasion, and so on.

Both the preform fabrication step and the fiber drawing steps are fully automated to achieve excellent and reproducible and results. By appropriate modifications, multimode and single-mode fibers can both be manufactured using this process.

4.11 NANOFIBERS

Recently, unclad optical fibers with diameters of 50 to 500 nm (0.05 to 0.5 μm) have been realized which are expected to find applications in microphotonics, sensing, and so on. Figure 4.23 shows a 450-nm unclad fiber tied into a knot placed on a human hair showing the relative size of the bare fiber. Such fibers currently show attenuation of a few dB/mm at 633-nm wavelength. Many applications are being

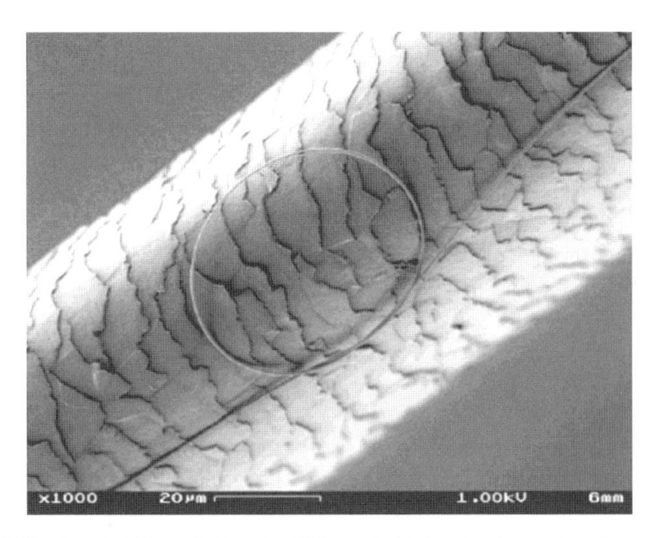

FIGURE 4.23 Unclad fiber of diameter 450 nm tied into a knot and placed on a human hair. (Adapted from Hecht, 2004.)

considered for such fibers. For example, the fibers can be bent tightly to form ring resonators. They are expected to exhibit very interesting nonlinear optical properties. The field penetrating the outside region of the fiber is very sensitive to the external region, so such fibers are expected to have very good sensing properties, especially for applications in biological sensing.

4.12 PLASTIC OPTICAL FIBERS

Whereas only glass fibers are used in long-distance communications, plastic fibers find applications in short-distance communications. They guide light using the same principle and could be cheaper where the distances are not great. Plastic optical fibers (POFs) are fibers made from plastic materials such as PMMA polymethyl methacrylate, polystyrene, polycarbonates, and fluorinated polymers. These fibers have the same advantages as those of glass optical fibers in terms of insensitivity to electromagnetic interference, small size and weight, low cost, and the potential to carry information at high rates of speed. The most important attribute of POFs is their large core diameters of around 1 mm compared to glass fibers with cores of 50 or 62.5 μm. Such a large diameter results in easy alignments at joints, for example. They are also more durable and flexible than glass fibers. In addition, they usually have a large index difference between core and cladding, resulting in greater light-gathering power.

Attenuation is an important parameter of an optical fiber. Losses in the fibers are typically high, with light transmission of only 10 to 25% over 100 m. Such losses are, of course, very large compared to those of silica fibers. The large losses in plastic fibers are due to such factors as Rayleigh scattering and intrinsic absorption of the material itself and of the impurities. Because of the high losses, these fibers are used only in short-distance (a few hundred meters) communication links.

Apart from short-distance communication applications, POFs are expected to find applications in many other areas, such as lighting for decorative and road signs; in museums, where the property of not transmitting ultraviolet radiation is very useful; and in image transmission, as in endoscopes and in sensing applications. The use of POFs in the industrial control and automotive fields appears to be increasing steadily. The main impetus for their use in the industrial control area is their immunity to electromagnetic interference caused by high-current devices, such as arc welders. POF use in automobiles for internal networking is expected to revolutionize this function.

4.13 FREE-SPACE OPTICS

Although using optical fibers is the preferred method of communication, as it has the highest levels of reliability and protection, in certain situations the distances are not large and installation of fibers could pose problems, such as densely populated city centers or due to geographic location. Thus, if communication is to be achieved from the nearest node to the end users, also known as the "last mile," it is possible to send information via free-space propagation of laser signals. Such a technique is referred to as free-space optics or optical wireless. In this case a narrow beam of light is launched at the transmitting end, and after propagating through open atmosphere, reaches the receiving end, where the signal is detected and processed. When laser beams propagate through open atmosphere, they diverge, and thus the amount of power reaching a receiver would be limited. In addition, atmospheric losses due to scattering, absorption, and so on, are in general large and it is not possible to use this type of propagation beyond distances of about 4 km. Such a system could be very easy to install, and new installations can be carried out very rapidly and cannot be jammed. Unlike microwave communication, permission is not required for use of the spectrum. The same technology that has been developed for fiber optic communication can be employed in this case to achieve very high capacities. Presently, commercial systems are available using either 750 or 1550-nm wavelengths. Optical wireless could also be an interesting solution in disaster recovery situations.

4.14 FIBER OPTIC CABLES

Like copper wires, optical fibers have to be converted to cable form for use in the field. A fiber optic cable consists of one or more optical fibers protected by either a loose tube or by coating with a buffer that fits tightly on the fiber. All fiber optic cables share some common characteristics. They all include various plastic coatings to protect the fiber and strength members, usually a high-strength *aramid* yarn to pull the cable without harming the fibers.

In *loose tube construction*, the fiber is placed within a plastic tube of larger inner diameter than the fiber diameter (Fig. 4.24*a*). The tube is usually filled with silicone gel to prevent moisture from seeping into the optical fiber. The loose tube design also protects the fiber from exterior mechanical forces acting on the cable. In this construction, since the fiber is placed loosely within the plastic tube and is longer than the tube itself, it does not suffer from mechanical perturbations when the cable is pulled or from environmental loading. This type of cable is used primarily in outside plant installations and is well suited for interoffice, long-haul, high-density, and distribution applications. Many such tubes can be assembled into a multifiber cable, or each tube may contain multiple fibers (6 to 12). Figure 4.25 shows a six-fiber cable

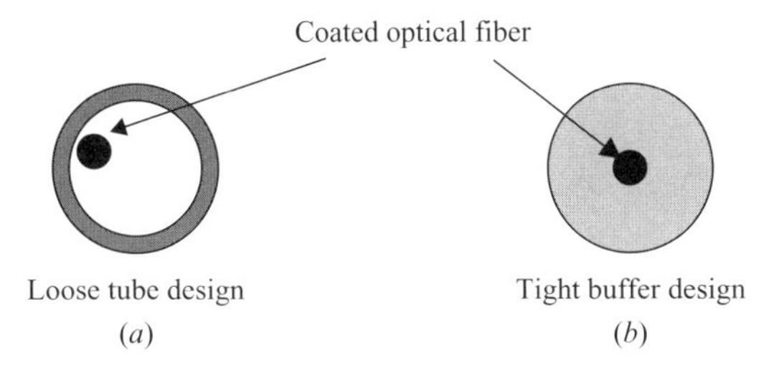

FIGURE 4.24 (*a*) Loose tube and (*b*) tight buffer designs.

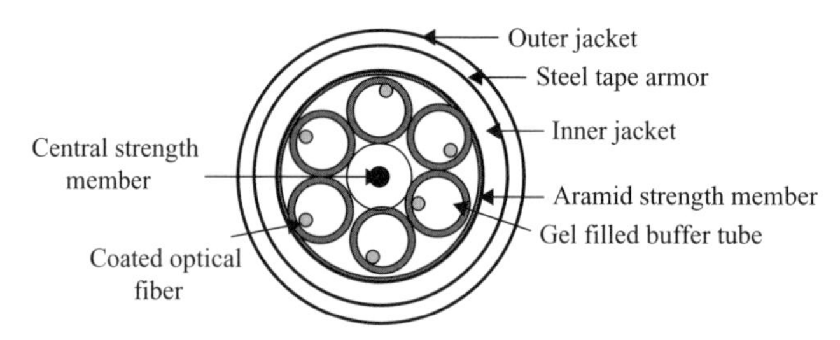

FIGURE 4.25 Fiber optic cable cross section consisting of six loose tube fiber optic cables with a central strength member.

in which the central member is either metallic or dielectric. Design identification of fibers is easy through the use of color coding. This design also permits easy drop of groups of fibers at intermediate locations. Armored cable designs utilizing a steel or dielectric central strength member may be used in duct, aerial, or buried applications. A cable can have from 2 to 264 fibers.

Tight buffer-coated fiber cable is made by direct extrusion of plastic over the fiber coating (Fig. 4.24*b*). This design is usually smaller in diameter, lighter in weight, and more flexible than loose tube construction. Such cables, used as patch chords within buildings or to interconnect instruments, can withstand much greater crush and impact forces without fiber breakage. Inside buildings, cables do not have to be very strong to protect fibers but have to meet all fire code provisions. The tight buffer design results in lower isolation of the fiber from stresses of temperature variation, which can induce microbending losses in the fiber.

In a third type of cable, *breakout cable*, a tightly buffered fiber is surrounded by aramid yarn and typically, a polyvinyl chloride jacket. This single-fiber unit is then covered by a common sheath to form a breakout cable, combining multiple-fiber flexibility with the strength of individually jacketed fibers. These cables are used in a variety of intrabuilding applications.

Figure 4.26 shows some common types of fiber optic cables.

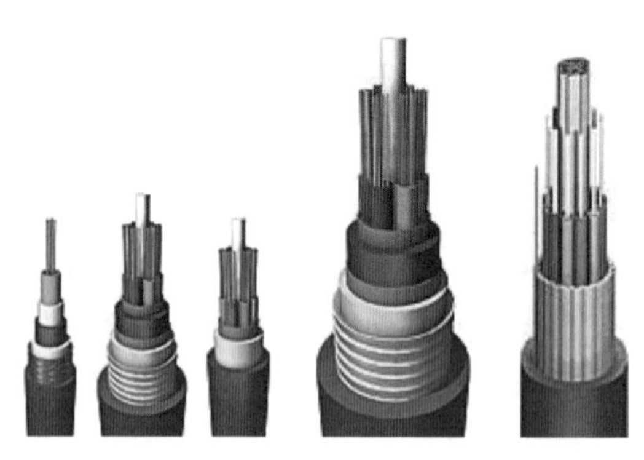

FIGURE 4.26 Common types of fiber cables. (From http://www.americantechsupply.com/alcoafiberopticcable.htm.)

CHAPTER 5

Loss in Optical Fibers

5.1 INTRODUCTION

Loss or attenuation and pulse dispersion represent the two characteristics of an optical fiber most important in determining the information-carrying capacity of a fiber optic communication system. In digital communication systems, information to be sent is first coded in the form of pulses, and these pulses of light are then transmitted from the transmitter to the receiver, where the information is decoded. A typical fiber optic communication system (Fig. 5.1) consists of a transmitter, which could be either a laser diode or a light-emitting diode, whose light is modulated by the signal and coupled into an optical fiber. Along the path of the optical fiber, there are splices, which are permanent joints between sections of fibers, and repeaters, which boost the signal and correct any distortion that may have accumulated along the path of the fiber. At the end of the link, the light is detected by a photodetector, which converts the optical signals to electrical signals, which are then processed electronically to retrieve the signal. The greater the number of optical pulses that can be sent per unit time and still be detectable and resolvable at the receiver end, the larger will be the transmission capacity of the system. A pulse of light sent into a fiber gets attenuated as it propagates through the fiber, and if the loss is large, there would not be enough light for the detector to separate the signal from the noise, and thus it cannot detect individual pulses. In addition to the attenuation, the pulse broadens in time as it propagates through the fiber. This phenomenon, known as *pulse dispersion*, is discussed in Chapters 6 and 7. Obviously, the lower the attenuation (and similarly, the lower the dispersion), the greater will be the required repeater spacing and therefore the lower will be the cost of the system. In this chapter we discuss briefly the various attenuation characteristics of an optical fiber.

5.2 THE DECIBEL UNIT

The attenuation of an optical beam is usually measured in the unit decibel (dB), used whenever the range of variation of a quantity is very large, so that with a manageable

Fiber Optic Essentials, By K. Thyagarajan and Ajoy Ghatak

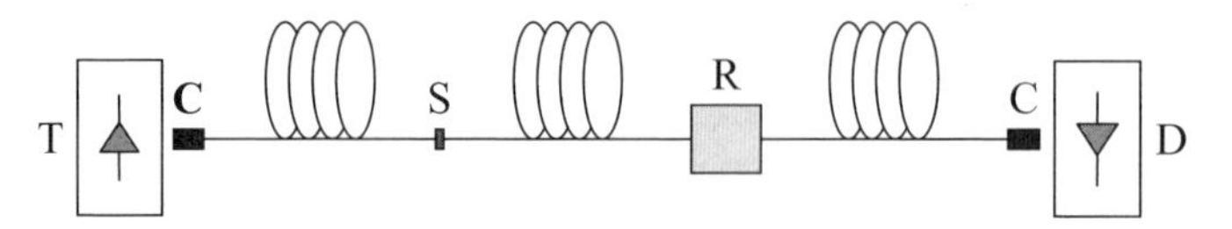

FIGURE 5.1 Typical fiber optic communication system. C, connector; S, splice; R, repeater; T, transmitter, D, detector.

range of numbers it is possible to describe the variation of the quantity. The decibel is related to the ratio of two power levels: for example, the ratio of output optical power from an optical fiber to the input optical power, or the ratio of output electrical power to the input electrical power of an electronic amplifier. If the ratio can vary over a very large range, it is convenient to use a logarithmic rather than a linear scale: If an input power P_1 results in an output power P_2, the loss in decibels is given by

$$\alpha = 10 \log_{10} \left(\frac{P_1}{P_2} \right) \tag{5.1}$$

Thus, if the output power is only half of the input power, the loss is $10 \log 2 \approx 3$ dB. Similarly, 20- and 10-dB losses will correspond to power reductions by factors of 100 and 10, respectively; every multiplication by a factor of 10 increases the value in decibels by 10. Thus, a loss of 60 dB would correspond to an output power, that is only one millionth of the input power. On the other hand, if 96% of the light is transmitted through the fiber, the loss is given by

$$\alpha = 10 \log_{10} \frac{1}{0.96} \approx 0.18\,\text{dB}$$

Similarly, in a typical fiber amplifier a power gain of 30 dB would imply power amplification by a factor of 1000.

If we consider an optical fiber of length L (km) and if for an input power of P_1, the output power is P_2, we define the loss coefficient of the fiber as

$$\alpha = \frac{10}{L} \log_{10} \frac{P_1}{P_2} \tag{5.2}$$

If for an input power of 1 mW, the power exiting the fiber after traveling 100 km is 10 μW, the loss coefficient of the fiber would be 0.2 dB/km. The loss coefficient in dB/km is related to the loss coefficient in km^{-1} by the formula

$$\alpha(\text{dB/km}) \approx 4.34\,\tilde{\alpha}\,(\text{km}^{-1})$$

The reason for measuring loss on a logarithmic scale is simple: If an optical fiber has a loss of 10 dB in traversing through 1 km (i.e., if the output power is only one-tenth of the input power in traversing through 1 km of the fiber), then in traversing through 3 km of the fiber, the output power will be one-thousandth of the input power (one-tenth for every kilometer traversed); that is, it will have a loss of 30 dB in traversing through 3 km. Thus, the loss in decibels just gets totaled! On the other hand, if the output power is only half of the input power, the loss is about 3 dB.

A common use of the decibel unit is in the description of the sound intensity level. The threshold of hearing corresponds to a sound pressure of 2×10^{-5} Pascals (Pa), referred to as 0 dB; 1 Pa corresponds to atmospheric pressure, usually abbreviated as 1 Pa. Intensity is proportional to square of pressure. Hence 60 dB sound level corresponds to an intensity which is million times the threshold of hearing. The corresponding pressure would be more by a factor of 1000 and thus corresponds to a pressure level of 2×10^{-2} Pa. Normal conversation corresponds to about a 60-dB sound level (or a pressure of about 2×10^{-2} Pa). The sound level at a rock concert corresponds to about 120 dB (pressure of about 20 Pa), and the eardrums start to feel pain at about 130 dB. For a person near a jet engine, the pressure can be so high that it can rupture the eardrums! Indeed, noise levels above 85 dB can cause damage to hearing over time.

Example 5.1 Calculation of losses become very easy using the decibel scale. For example, if we have a 80-km fiber link (with a loss of 0.25 dB/km) with five connectors in its path, and if each connector has a loss of 2 dB, the total loss will be (0.25 dB/km)(80 km) + (5)(2 dB) = 30 dB. Thus, the power will decrease by a factor of 1000.

Example 5.2 Let us assume that the power of a 10-mW laser beam decreases to 20 μW after traversing through 40 km of an optical fiber. The attenuation of the fiber is therefore $[10 \log(500)]/40 \approx 0.67$ dB/km.

5.3 THE dBm

It is also possible to specify power levels using the decibel unit by using a reference power. Thus, if we wish to choose the reference power as 1 milliwatt (mW; one-thousandth of a watt), the logarithmic unit of power with reference to 1 mW is referred to as dBm:

$$P(\text{dBm}) = 10 \log_{10} P(\text{mW}) \tag{5.3}$$

Thus,

$$\begin{aligned} &1\,\text{mW} \Leftrightarrow 0\,\text{dBm} \\ &0.2\,\text{W} = 200\,\text{mW} \Leftrightarrow \approx 23\,\text{dBm} \end{aligned}$$

Similarly, power levels of 10, 100, and 1000 mW (=1 W) would correspond to +10, +20, and +30 dBm, respectively. Further, power levels of 1 microwatt (μW; one-millionth of a watt) and 1 nanowatt (nW; one-billionth of a watt), would correspond to −30 and −60 dBm, respectively.

Using the dBm scale, Eq. (5.1) becomes

$$\alpha = P_{\text{input}}(\text{dBm}) - P_{\text{output}}(\text{dBm})$$

or

$$P_{\text{output}}(\text{dBm}) = P_{\text{input}}(\text{dBm}) - \alpha\,(\text{dB}) \tag{5.4}$$

Thus, when we use the dBm scale, the power calculation becomes very easy. For example, if a 10-mW pulse (= 10 dBm) undergoes a loss of 30 dB, the output power will be $10 - 30 = -20$ dBm (= 0.01 mW).

Example 5.3 Consider a 5-mW laser beam passing through a 40-km fiber link of loss 0.5 dB/km. The total loss is 20 dB. Since the input power is 6.99 dBm, the power at the output would be -13.01 dBm, which is equal to 0.05 mW.

Example 5.4 Let us assume that the power of a 10-mW (= 10 dBm) laser decreases to 10 μW (-20 dBm) after traversing through 30 km of an optical fiber. The total loss is 30 dB and the attenuation of the fiber is therefore 1 dB/km.

Example 5.5 Let the input power of a fiber link be 1 mW (= 0 dBm). If there are two connectors that have a loss of 1 dB per connector, the total connector loss is 2 dB. In addition, if there are four splices (joints) and there is a loss of 0.5 dB/splice, the total splice loss will be 2 dB. Further, if the fiber length is 32 km, characterized by a loss of 0.5 dB/km, the fiber loss will be 16 dB. Thus, the total loss will be $2 + 2 + 16 = 20$ dB, and the power received will be -20 dBm (= 10 μW).

5.4 LOSS MECHANISM IN OPTICAL FIBERS

When light propagates through any medium, even the purest materials, it suffers loss, due to various mechanisms: scattering and absorption, among others, caused by the atoms and molecules that form the material. As can be seen from Figure 5.2, which shows the evolution of losses in glass from ancient times, in 1966 the most transparent glass available had a loss of 1000 dB/km, due primarily to trace amounts of impurities present in the glass. A loss of about 1000 dB/km (or equivalently, 1 dB/m) implies that for every 10 m the power will fall by a factor of 10. Thus, after propagating through 1 km of such a fiber, the output power will be 10^{-100} of the input power; this value is, for all practical purposes, zero (i.e., almost no light will emerge from the output).

As discussed in Chapter 4, in 1966 Kao and Hockham (1966) first suggested the use of optical fibers for communication and pointed out that for the optical fiber to be a tenable communication medium, losses (in the optical fiber) should be less than 20 dB/km. This suggestion triggered the beginning of serious research into removing the small amount of impurities present in the glass and developing low-loss optical fibers. Indeed, since 1966, there has been a global effort to purify silica, and in 1970 there was a major breakthrough: The first low-loss glass optical fibers were fabricated by Maurer, Keck, and Schultz (at Corning Glass in the United States) with the optical fibers fabricated having a loss of about 17 dB/km at a wavelength of 633 nm. This would imply that the power will fall by a factor of 10 in traversing

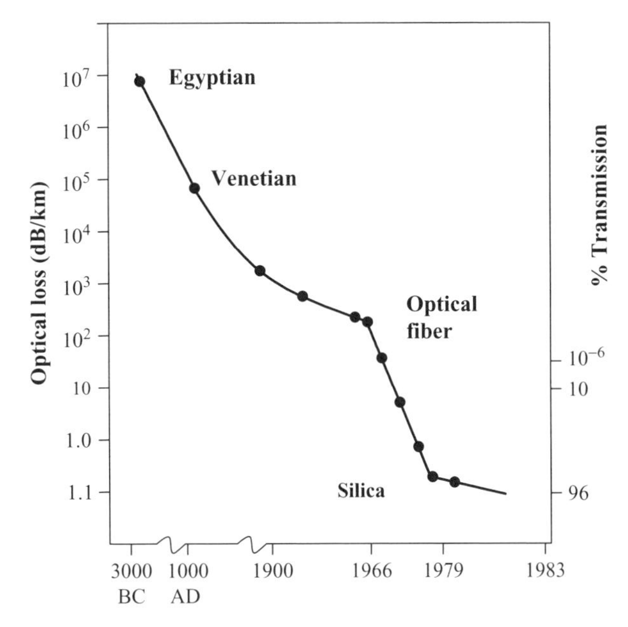

FIGURE 5.2 The evolution of losses in glass from ancient times. The most transparent glass available in 1966 had a loss of about 1000 dB/km, due primarily to trace amounts of impurities present in the glass. (Adapted from Nagel, 1989. Copyright © 1989 IEEE.)

through approximately a 600-m length of fiber. Since then, the technology has been improving continuously and by the late 1980s, commercially available optical fibers had losses of less than about 0.25 dB/km at a wavelength of about 1550 nm (a loss of 0.25 dB/km implies that the power will fall by a factor of 10 after propagating through a 40-km length of the optical fiber).

Figure 5.3 shows a typical dependence of fiber attenuation (i.e., loss coefficient per unit length) as a function of wavelength of a typical silica optical fiber. It may be seen that the loss is about 0.25 dB/km at a wavelength of about 1550 nm. The modified chemical vapor deposition (MCVD) process (discussed in Chapter 4) allows us to fabricate such fibers with very low losses. The losses in optical fibers are caused by various mechanisms, such as Rayleigh scattering, absorption due to metallic impurities and water in the fiber, and intrinsic absorption by the silica molecule itself.

Rayleigh scattering is a basic mechanism by which light gets scattered by very small inhomogeneities as it propagates through any medium. Rayleigh scattering loss is wavelength dependent and is such that shorter wavelengths scatter more than longer wavelengths with the loss proportional to λ^{-4}, where λ is the optical wavelength. It is this phenomenon that is responsible for the blue color of the sky. As sunlight passes through the atmosphere, the component corresponding to the blue color gets scattered more than the component corresponding to the red color (since blue wavelengths are shorter than red wavelengths). Thus, more blue than red reaches our eye from the

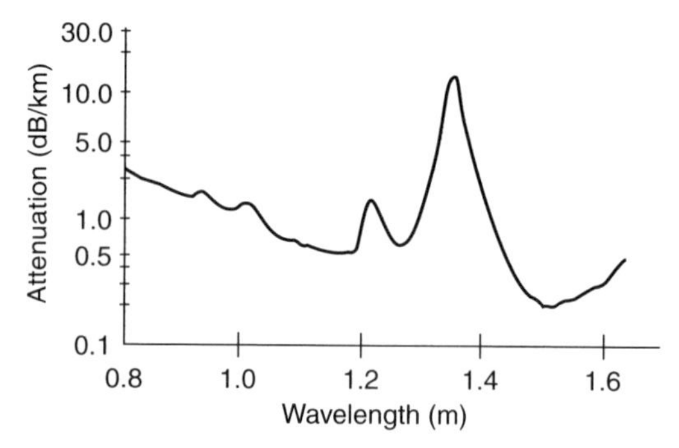

FIGURE 5.3 Typical wavelength dependence of attenuation for a silica fiber. Notice that the lowest attenuation occurs at 1550 nm. (Adapted from Miya et al., 1979.)

sky and the sky looks blue. This is also the reason that both a rising and a setting sun appear red to our eyes.

Rayleigh scattering causes attenuation of optical signals as they propagate through an optical fiber. Very small inhomogeneities present in the fiber scatter light out of the fiber, leading to loss. It is Rayleigh scattering that makes the fiber visible in Fig. 4.4. In fact, this loss mechanism determines the ultimate loss of optical fibers. Since Rayleigh scattering loss decreases with increased in wavelength, optical fibers operating at higher wavelengths are expected to have lower losses if all other loss mechanisms are eliminated.

Apart from Rayleigh scattering loss, any impurities present in an optical fiber would also cause the absorption of propagating light and thus contribute to loss. Primary impurities include such metallic ions as copper, chromium, iron, and nickel. Impurity levels of 1 part in a billion could cause increase in attenuation of 1 dB/km in the near-infrared region. Such impurities can be reduced to acceptable levels by using vapor-phase oxidation methods. Apart from impurity metal ions, one of the major contributors to loss is the presence of water dissolved in the glass. An impurity level of just 1 part per million (1 ppm) of water can cause a loss of 4 dB/km at 1380 nm. This shows the high level of purity that is required to achieve low-loss optical fibers.

Figure 5.3 shows the attenuation spectrum of a typical optical fiber. The primary reason that the loss coefficient decreases up to about 1550 nm is the Rayleigh scattering loss. The two absorption peaks around 1240 and 1380 nm are due primarily to traces of water and traces of metallic ions. If these impurities are removed completely, the two absorption peaks will disappear (see Fig. 5.4) and we will have very low loss in the entire range of wavelength from 1250 to 1650 nm. Such fibers are now available commercially (Fig. 5.5). For wavelengths longer than about 1650 nm, the loss increases, due to the absorption of infrared light by the silica molecules themselves. As this is an intrinsic property of silica, no amount of purification can remove this infrared absorption tail.

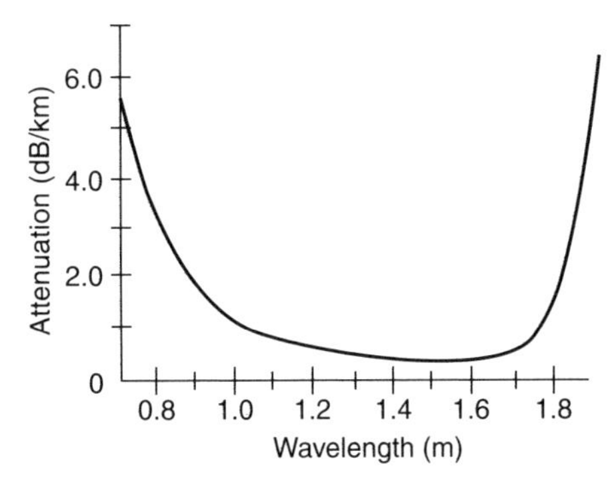

FIGURE 5.4 Loss spectrum of an ultimately low water content in an optical fiber. Note that the low-loss window extends from 1250 to 1650 nm (about 50 THz), and such fibers are now available commercially. (Adapted Moriyama et al., 1980.)

As we can see from Fig. 5.3, there are two windows at which loss attains its minimum value. The first window is around 1300 nm (with a typical loss coefficient of less than 1 dB/km), where fortunately (as we will see in Chapter 6), the material dispersion is negligible. However, the loss attains its absolute minimum value of about 0.16 dB/km when the wavelength is around 1550 nm, as a consequence of which the distance between two consecutive repeaters (used for amplifying and reshaping the attenuated signals) could be increased significantly. Furthermore, the 1550-nm window has become extremely important in view of the availability of erbium-doped fiber amplifiers (see Chapter 9).

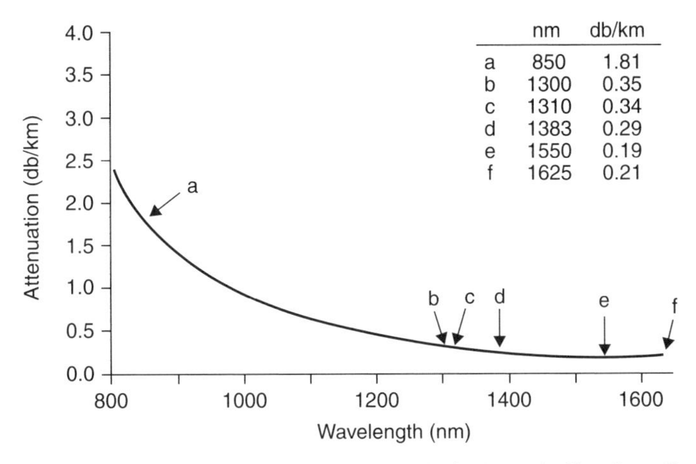

FIGURE 5.5 Loss spectrum of a low-water-peak single-mode fiber from Corning.

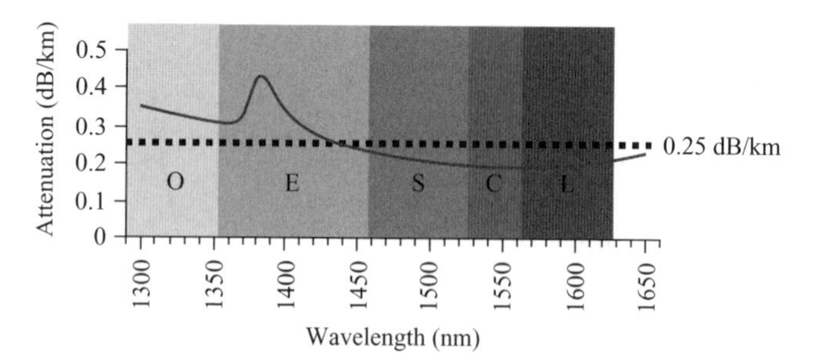

FIGURE 5.6 Loss spectrum of LEAF fiber from Corning. The low-loss spectrum is divided into various bands, which are given in Table 5.1.

TABLE 5.1 Wavelength Bands Used in Optical Fiber Communication Systems

Wavelength Band	Wavelength Region (nm)
O-band (old band)	1260–1360
E-band (extended band)	1360–1460
S-band (short band)	1460–1530
C-band (conventional band)	1530–1565
L-band (long band)	1565–1625

Figure 5.6 shows the attenuation spectrum of a commercial fiber (called LEAF, for "large effective area fiber") from Corning superimposed on the various wavelength bands of operation that can be used with such low-loss fibers. The wavelength ranges corresponding to various wavelength bands are listed in Table 5.1. Current fiber optic communication systems use primarily the C- and the L-bands. The coarse wavelength-division-multiplexing (CWDM) scheme utilizes wavelength channels spaced 20 nm apart. Due to the large channel spacing, the wavelength of a transmitter need not be very precise, and thus uncooled laser diodes can be used. Such a system was developed for metropolitan applications in which cost is a very important factor.

Other Loss Mechanisms

Apart from intrinsic loss mechanisms, any bend in a fiber path would result in added loss, referred to as *bending loss* (Fig. 5.7). This additional loss depends on such fiber parameters as the core–cladding index difference, the core radius, and the wavelength of operation. Thus, in the case of single-mode fibers, the bend-induced loss depends on how tightly the mode is bound to the core. Optical fibers with a larger numerical aperture (a larger difference between the core and cladding refractive indices) would have tighter confinement and hence would suffer from lower bend-induced loss.

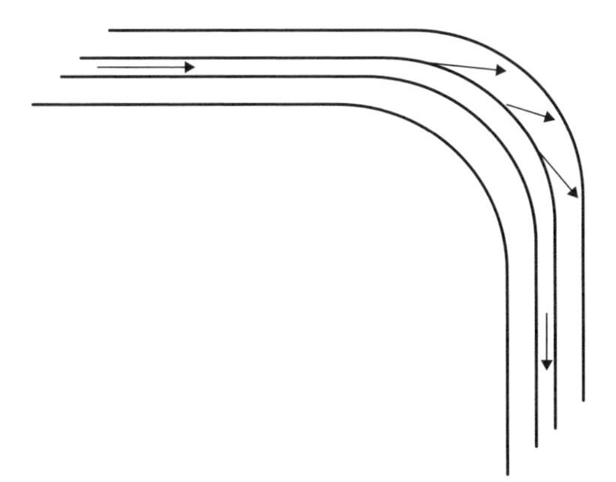

FIGURE 5.7 When a fiber is bent, light is lost from the outer surface of the bend. Optical fibers need to satisfy stringent criteria in bend-induced loss.

In the case of standard single-mode fibers, international standards for fiber optic systems require that the additional bend-induced loss due to 100 turns at a radius of 30 mm be less than 0.5 dB. Such bends are found in splice boxes in repeater stations, where extra fiber lengths are provisioned for any repairwork that may have to be carried out later in the life of the system. To prevent excessive bend-induced loss, all optical fibers are required to have a minimum bend radius that should not be exceeded.

Bend losses have become especially significant with the installation of optical fibers in the last mile of "fiber to the home" (FTTH) services. In such applications, optical fibers would be subjected to a greater degree of bending in cabinets, in laying them within buildings, and so on. This has led to much more stringent requirements regarding bend-induced loss in optical fibers. For this, special designs of optical fibers with greater bend resistance have been developed. This added feature has been achieved while maintaining a large mode field for proper splicing with standard fibers, low loss at the water peak, and chromatic dispersion consistent with standard single-mode fibers. Such fibers are specified by the ITU-T standard G.652D. An example of such a fiber is the AllWave FLEX ZWP fiber from OFS in the United States. Unlike standard fibers, which would have enhanced bend losses at these wavelengths, these fibers also show very good performance up to wavelengths of 1625 nm for bend radii of even 20 mm. Figure 5.8 shows the bend loss induced in a typical bend-insensitive fiber compared to standard G.652 single-mode fiber and another type of bend-insensitive fiber.

If we imagine the fiber to be laid on a rough surface and if pressure is applied from the top, the fiber would have microscopic random bends of small radii of curvature all along its length (Fig. 5.9), referred to as *microbending*. This introduces additional loss in the fiber, due to coupling of light from the core to the cladding, from where the light is lost. Such loss can be caused while cabling the fiber, from irregularities during fiber manufacture, or by environmental stresses, which can cause differential

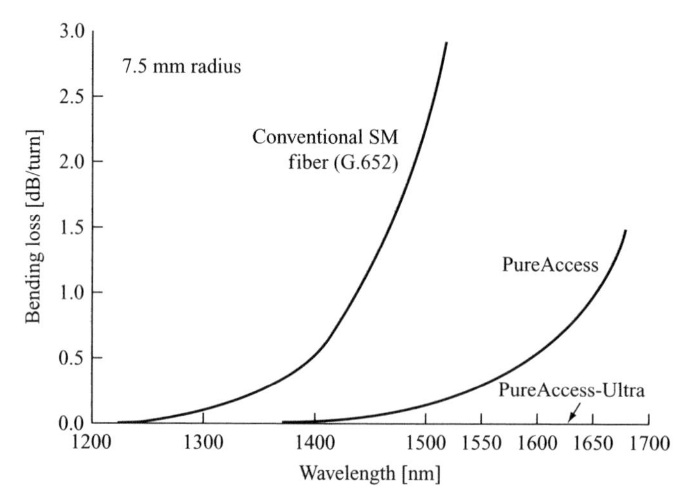

FIGURE 5.8 Comparison of wavelength-dependent loss of a standard single-mode fiber with special fiber designs exhibiting reduced bend loss. (Adapted from Sakabe et al., 2005.)

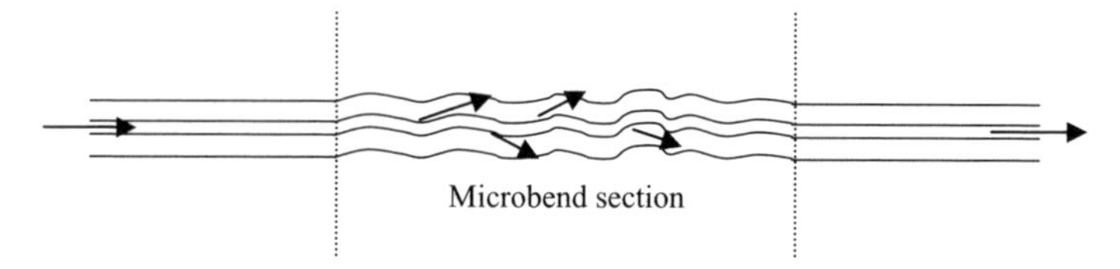

FIGURE 5.9 Microscopic random bends of small radii of curvature along an optical fiber induce loss in an optical fiber, referred to as microbend loss.

expansion or contraction. Microbend-induced loss can also be used to sense external perturbations (discussed in Chapter 14).

In an actual fiber optic link, optical fibers are spliced (fused) or given temporary connections via connectors. At these points additional loss can take place, due to small misalignments between the cores of the fiber or to the two fibers not being identical. Such losses, termed *splice losses* and *connector losses*, are discussed in Chapter 12.

5.5 DEMONSTRATION OF RAYLEIGH SCATTERING IN AN OPTICAL FIBER

Rayleigh scattering is responsible for the blue color of the sky and the red color of the setting sun. It is very interesting to demonstrate the wavelength dependence of Rayleigh scattering using a long optical fiber. Couple white light from a lamp such as a tungsten halogen lamp emitting white light into an approximately 1-km-long multimode optical fiber and notice the color of the light when looking into the output.

Cut the fiber, leaving about 1 m from its input end, and repeat the experiment with this 1 m of the fiber. You would see that in the former case, the emerging light looks reddish, whereas in the latter case it looks white. This difference is due to the decrease of loss with increase in wavelength due to Rayleigh scattering; light wavelengths toward the blue region have suffered greater scattering out of the fiber than those of the red region. Thus, although at the input end all wavelengths are coupled, there is more power in the red part at the output, giving it a reddish color.

5.6 OPTICAL TIME-DOMAIN REFLECTOMETER

An optical time-domain reflectometer (OTDR), which is like a radar, is a very versatile instrument that is capable of locating breaks or other perturbations, such as sharp bends, connectors, and splice positions, and their losses, as well as loss distributed along the length of an optical fiber, by having access to only one end of the fiber. It is very useful in characterizing optical fiber links after they are laid and in monitoring any changes in loss that can take place with time.

Basic Principles

In an OTDR an intense light pulse is launched from one end of an optical fiber under test through a beamsplitter or a directional coupler (Fig. 5.10). As the light pulse propagates through the fiber, it undergoes Rayleigh scattering in all directions and suffers partial reflection at connectors and splices. The amount of light scattered depends on the inhomogeneities present in the fiber and on the power carried by the light at that point. A tiny fraction couples back into the fiber, and propagates in the reverse direction along the fiber, and exits from the input end of the fiber. The OTDR shows the time variation of the light energy exiting the input end of the fiber. The backscattered light power exiting the input end at any time is produced by a certain section of the fiber located at a certain distance from the input end. Thus, from the

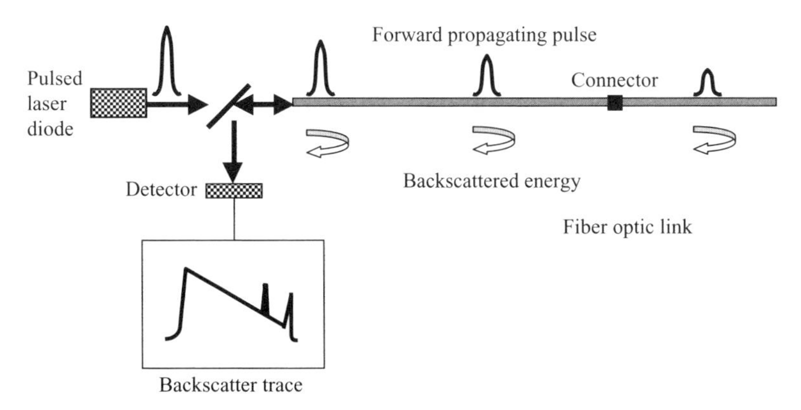

FIGURE 5.10 Schematic diagram of an OTDR. The first peak in the backscatter trace is generated by the connector; the second is due to reflection from the fiber end.

time variation of the backscattered light power, it is possible to estimate the variation of the power within the fiber and hence the loss along the fiber length. Whenever there is a break or a connector, discontinuity in the fiber would result in a back reflection, which would appear as a spike in the OTDR trace (Fig. 5.10). The location of the spike in the trace can be used to estimate the position of the connector or break in the fiber.

To better understand the operation, imagine a pulse of light entering the fiber at a certain instant of time. As soon as the pulse is incident on the fiber, a small fraction (about 4%) is reflected from the input end, and a major fraction of the remaining 96% enters the fiber. As this pulse propagates through the fiber, it gets scattered and the light scattered in the backward direction exits from the input end. Let us consider the light that is exiting at a certain time, say 10 μs later. This energy must have been scattered by a certain region of the fiber which is at such a distance from the input end that light would have taken 5 μs to reach from the input end and another 5 μs to return to the input end. Since the speed of light in an optical fiber is about 200,000 km/s, the distance from the input end from where the light must have got scattered is about $(200{,}000)(0.000005) = 1$ km. Similarly, the energy exiting from the input end 20 μs later would have gotten scattered from the region of fiber 2 km from the input end. Thus, there is a direct correlation between the time light exits from the input end and the distance from which it has been scattered.

Now let us estimate the typical power level that would be returning. If the fiber is assumed to have a loss of 0.3 dB/km, then after every kilometer of propagation, the power in the pulse decreases by 0.3 dB, and if a certain fraction of this is backscattered, the power of the backscattered light will decrease an additional 0.3 dB in propagating over 1 km of the fiber so as to reach the input end. Thus, if we launch 10 mW of power, after 1 km the power in the pulse would be about 9.33 mW (loss of 0.3 dB). Typically, a fraction of 0.0001 of the power at any given position is the backscattered power, and thus the backscattered power at a distance of 1 km from the input end would be 0.933 μW. This backscattered power will become 0.871 μW (again a loss of 0.3 dB) by the time it reaches the input end. Similar power returning from a distance of 2 km from the fiber input end would be 0.759 μW. Thus, the power returning from the fiber decreases with time and the OTDR essentially shows this trace on a screen.

It is possible to estimate the attenuation coefficient of the fiber from this trace. In the example above, the power levels measured at 10 μs (corresponding to scattering from a distance of 1 km from the input end) and 20 μs (corresponding to scattering from a distance of 2 km from the input end) would be 0.871 μW and 0.759 μW, respectively. From these two values we can estimate the loss to be 0.6 dB, which is the loss in propagating through 2 kilometer of the fiber (1 km in the forward direction and 1 km in the reverse direction). Hence, the loss coefficient is 0.3 dB/km. In fact, the slope of the trace is representative of the attenuation coefficient of the fiber. The larger the attenuation coefficient, the larger will be the variation in power of the pulse as it propagates through the fiber, and the larger will be the slope of the trace. Similarly, a smaller attenuation coefficient would lead to a smaller slope.

Typically, a good splice joint produces almost no reflection but introduces some loss. Hence, the power in the light pulse has a sudden drop as it crosses a splice, and this is reflected as the drop shown in Fig. 5.11. If the fiber before and after the splice

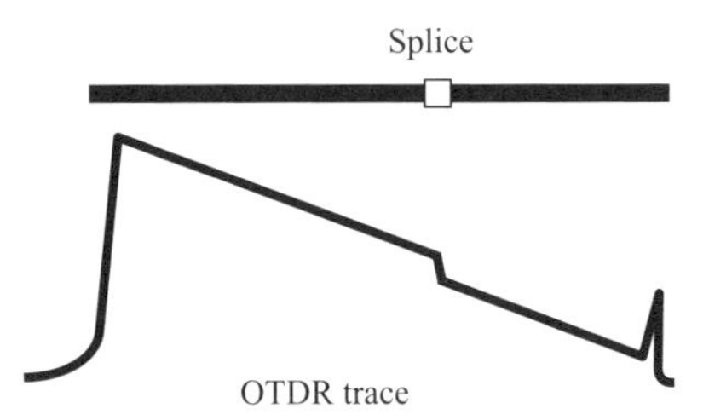

FIGURE 5.11 The OTDR trace shows a drop in backscattered power when the pulse crosses a splice. Good splices do not produce a reflection.

is identical, the slope of the curve before and after the drop would be the same. From the drop in the trace it is possible to estimate the splice loss. The position of the drop would also indicate the position of the splice in the link.

The backscattered energy depends on the peak power and on the pulse duration of the pulse that was launched. Typical pulse widths in OTDRs are in the range of 10 ns to a few microseconds. Larger pulse widths would lead to lower resolution in the of position measurement along the length. A pulse of duration 0.1 μs would have a physical length of $10^{-7} \times (2 \times 10^8) = 20$ m (duration of the pulse multiplied by the speed of the pulse in the fiber). Thus, using such a pulse it would not be possible to isolate events such as breaks, splices, and connectors that lie closer than about 20 m. Since the backscattered power received is very small (less than a nanowatt), a very sensitive avalanche photodetector (APD) is used. This instrument measures the backscattered power a number of times by sending a number of identical pulses through the fiber, which are then averaged to produce a trace with much less noise (Fig. 5.12).

Sometimes in an OTDR trace one can see features such as that shown in Fig. 5.13. Such an event is termed a *gainer*. This can happen when two different fiber types are

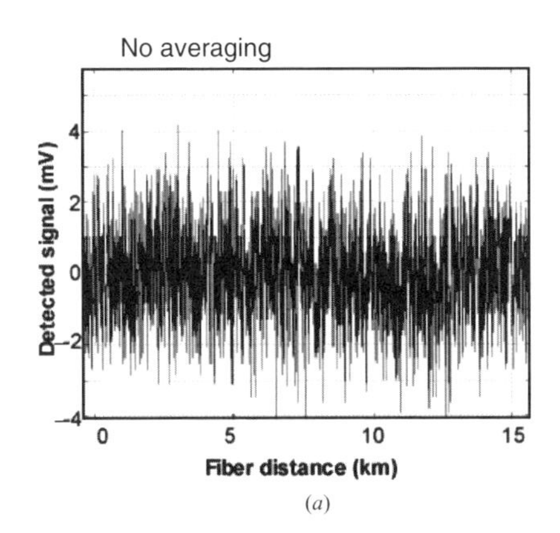

(*a*)

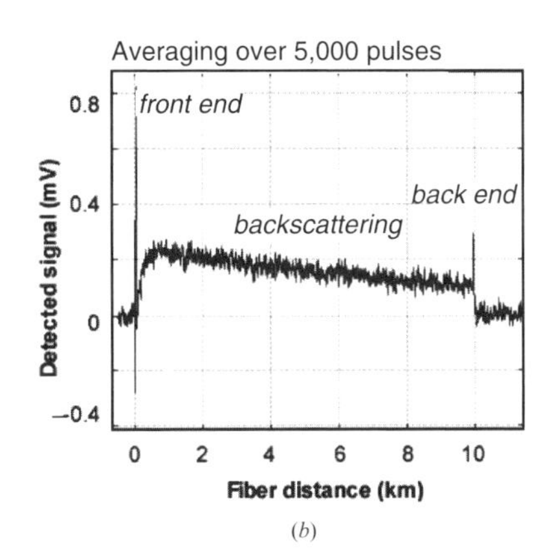

(*b*)

FIGURE 5.12 (*a*) Detection signal from a 10-km-long single-mode fiber when a laser pulse with a peak power of 13 mW and pulse duration of 100 ns is launched; (*b*) the same signal averaged over 50,000 sweeps. The trace shows clearly the reflection from both ends of the fiber and the fiber attenuation. (Adapted from Lucas-Lectin et al., 2005.)

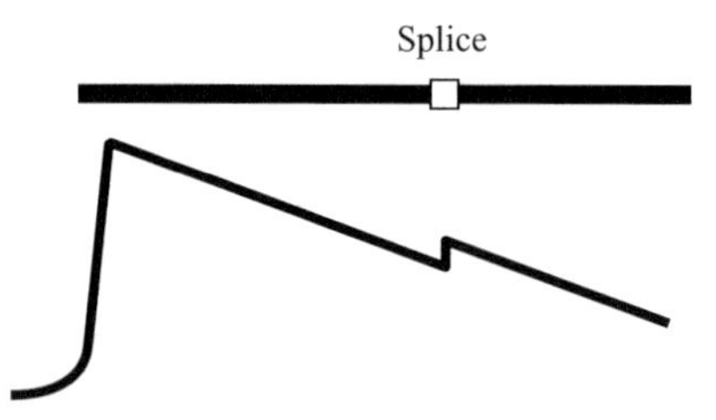

FIGURE 5.13 When the fiber type changes across a splice or joint, it is possible to observe an increase in the backscattered power.

joined. The power that is backscattered into the fiber depends on the back-scattering coefficient at that point and on the fraction of the scattered power that can be coupled back into the fiber. The latter quantity depends on the numerical aperture of the fiber; the higher the refractive index difference between the core and cladding, the larger the numerical aperture and the larger would be the fraction of the power that couples back into the fiber. Thus, if the second fiber shown in Fig. 5.13 has a higher numerical aperture, it is possible that although the power level drops as the light pulse enters the second fiber, since the fractional power that couples back into the fiber is larger for the second fiber, the backscattered power may increase when the pulse passes from the first fiber to the second. Figure 5.14 shows the OTDR traces of a dispersion-compensating fiber (with high numerical aperture), followed by a standard single-mode fiber and the OTDR trace from the reverse direction. The drop and gain in the two cases are very clear. In such a case, by deriving the OTDR trace by launching the pulse from each direction, it is possible to estimate the splice loss at the joint.

The following are the principal characteristics of an OTDR:

1. *Wavelength of operation:* OTDRs use primarily the three wavelength bands 850, 1310, and 1550 nm. Sometimes, even 1625 nm is used for testing of fiber systems carrying live traffic, so that the wavelength does not interfere with the signals, which are usually in the 1550-nm window. Typically, the power levels used are in the range 10 mW to watts.
2. *Dynamic range:* This is a measure of the maximum power excursion range that the instrument can measure. This, in turn limits the maximum distance that an OTDR can access. The larger the dynamic range, the greater the signal-to-noise ratio and the better the detection capabilities will be.
3. *Dead zone:* When a laser pulse is launched into a fiber, it first suffers a strong reflection from the input end of the fiber. This reflection is much stronger than the backscattered radiation (more than 4000 times stronger) that the OTDR has been designed to detect, and this high power saturates the detector, which takes a finite time to recover. During this time the pulse is propagating along the length of the fiber, and any event in the fiber that occurs within this time will not be detected by the OTDR. This distance, referred to as the *dead zone*, is typically, about 10 m.
4. *Resolution:* The display resolution determines the minimum resolvable variation in power and hence loss. The loss resolution is typically 0.01 dB.

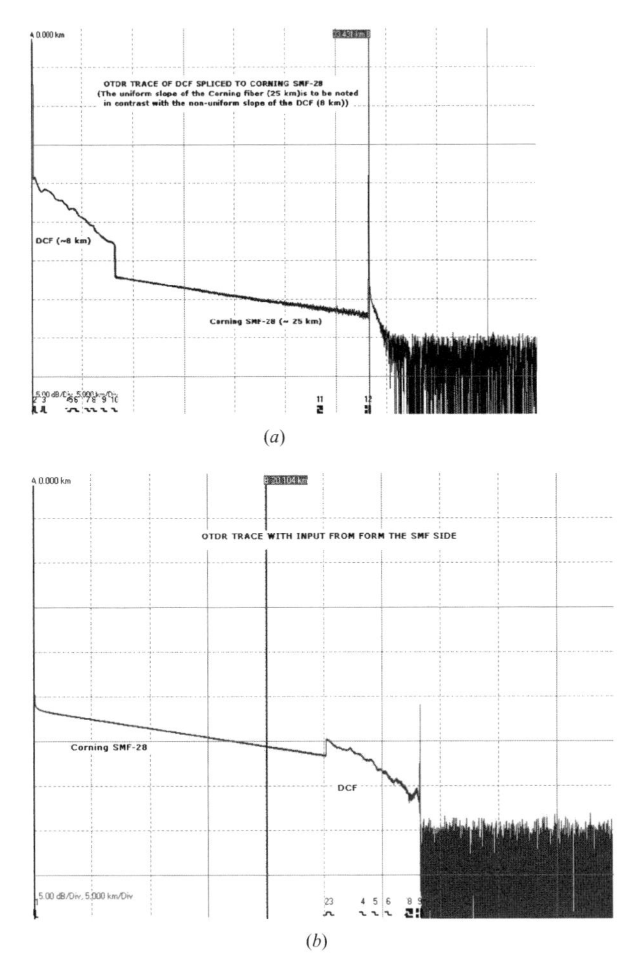

FIGURE 5.14 OTDR traces of (*a*) dispersion-compensating fiber (DCF) followed by single-mode fiber (SMF) and (*b*) SMF followed by DCF. Note that in the second case the backscattered power increases at the splice point.

5. *Distance accuracy:* The distance accuracy depends on accurate knowledge of the speed of the pulse and the pulse width. The speed of the pulse is determined by the refractive index of the fiber, which is usually provided by the fiber manufacturer. The speed of the light pulse, and hence the index, can also be estimated by measuring a known length of the fiber. It is interesting to point out that fibers can be placed loosely within the cable or wound in the form of a helix, causing the fiber length to be greater than the cable length. The shorter the pulse width, the better the resolution will be. At the same time, shorter pulses lead to smaller backscattered power and thus limit the range of an OTDR.

CHAPTER 6

Pulse Dispersion in Multimode Optical Fibers

6.1 INTRODUCTION

As discussed in Chapter 3, in digital communication systems, information to be sent is first coded in the form of pulses and then these pulses of light are transmitted from the transmitter to the receiver, where the information is decoded (Fig. 6.1). Further, the two most important characteristics of an optical fiber that determine its information-carrying capacity are (1) loss or attenuation (discussed in Chapter 5) and (2) pulse dispersion. As we discuss in this chapter, a pulse of light sent into an optical fiber broadens in time as it propagates through the fiber; this phenomenon is known as *pulse dispersion* (Fig. 6.2). The larger the number of pulses that can be sent per unit time and still be resolvable at the receiver end, the larger will be the transmission capacity of the system. If the pulse dispersion is large, then in order that consecutive output pulses be resolvable by the receiver so that information can be retrieved, the time interval between adjacent pulses has to be greater than a certain minimum value. This would then limit the number of pulses that can be sent per unit time and hence the information capacity of the system. Hence, the smaller the pulse dispersion, the greater will be the information-carrying capacity of the system. Thus, in communication applications one always tries to reduce pulse dispersion.

As a pulse propagates through the fiber, its temporal broadening occurs primarily because of the following mechanisms:

1. In multimode fibers, different rays take different times to propagate through a given length of the fiber; we discuss this for a step-index fiber and for a parabolic index fiber in this and the following sections. In the language of wave optics, this is known as *intermodal dispersion* because it arises due to different modes traveling with different velocities.
2. Any given light source emits over a range of wavelengths, and because of the intrinsic property of the material of the fiber, different wavelengths take different amounts of time to propagate along the same path. This is known as *material dispersion* and obviously is present in both single-mode and multimode fibers.

Fiber Optic Essentials, By K. Thyagarajan and Ajoy Ghatak

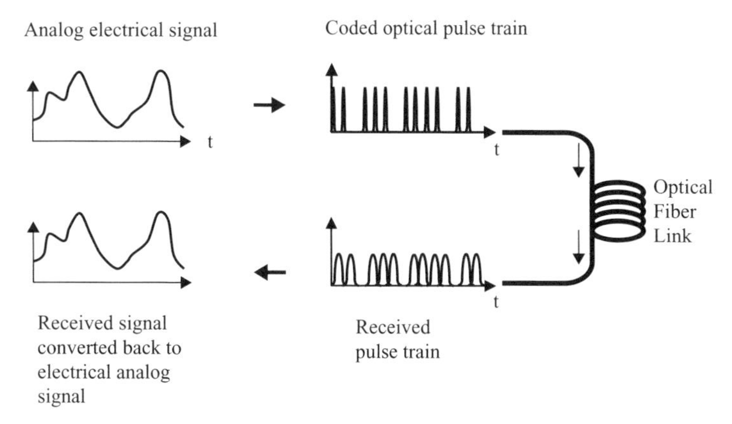

FIGURE 6.1 In digital communication systems, information to be sent is first coded in the form of pulses, and then these pulses of light are transmitted from the transmitter to the receiver, where the information is decoded.

FIGURE 6.2 A pulse of light sent into a fiber broadens in time as it propagates through the fiber, a phenomenon known as pulse dispersion.

3. In single-mode fibers, since there is only one mode, there is no intermodal dispersion. However, we have another mechanism that leads to dispersion, referred to as *waveguide dispersion*. Physically, this arises due to the fact that the spot size (of the fundamental mode) itself depends on the wavelength. Obviously, waveguide dispersion is present in multimode fibers also, but the effect is very small and can be neglected.
4. In the case of single-mode fibers operating at high bit rates, it is also important to consider the phenomenon of *polarization mode dispersion*; this is discussed in Chapter 7.

In the following sections we discuss the various dispersion mechanisms in multimode fibers in detail, and in Chapter 7 we discuss dispersion in single-mode fibers.

6.2 MULTIMODE FIBERS

A broad class of *multimode* graded-index fibers can be described by the following power-law refractive index distribution:

$$n^2(r) = \begin{cases} n_1^2\left[1 - 2\Delta\left(\dfrac{r}{a}\right)^q\right] & 0 < r < a \\ n_2^2 = n_1^2(1 - 2\Delta) & r > a \end{cases} \tag{6.1}$$

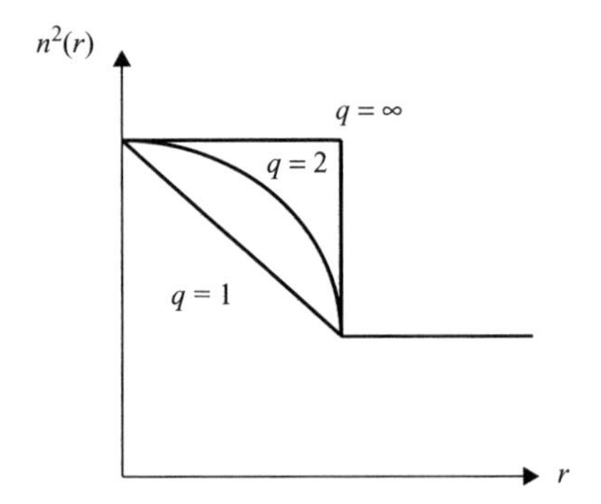

FIGURE 6.3 Refractive index variations for a power-law profile.

where r corresponds to a cylindrical radial coordinate (i.e., distance from the axis of the fiber), n_1 represents the value of the refractive index on the axis (i.e., at $r = 0$), and n_2 represents the refractive index of the cladding; $q = 1$, $q = 2$, and $q = \infty$ correspond to the linear, parabolic, and step-index profiles, respectively (Fig. 6.3). Further,

$$\Delta \equiv \frac{n_1^2 - n_2^2}{2n_1^2} = \frac{(\mathrm{NA})^2}{2n_1^2} \tag{6.2}$$

where NA represents the numerical aperture of the fiber and is given by

$$\mathrm{NA} = \sqrt{n_1^2 - n_2^2} \tag{6.3}$$

When $n_1 \approx n_2$, that is, when $\Delta \ll 1$ (as is true for most silica fibers),

$$\Delta = \frac{n_1 - n_2}{n_1} \frac{n_1 + n_2}{2n_1} \approx \frac{n_1 - n_2}{n_2} \approx \frac{n_1 - n_2}{n_1} \tag{6.4}$$

Equation (6.1) describes what is usually referred to as a *power-law profile*, which gives an accurate description of the refractive-index variation in most multimode fibers. The total number of modes in a highly multimoded graded-index optical fiber characterized by Eq. (6.1) is given approximately by

$$N \approx \frac{q}{2(2+q)} V^2 \tag{6.5}$$

Thus, a parabolic index ($q = 2$) fiber with $V = 10$ will support approximately 25 modes. Similarly, a step-index ($q = \infty$) fiber with $V = 10$ will support approximately 50 modes. A typical multimode fiber would have a core diameter of 50 μm and a Δ value of 0.02, giving a V value of about 24 at 1310 nm. For this value of V, a parabolic index fiber would support about 144 modes.

When the fiber supports such a large number of modes, the description of light propagation through such fibers using rays should give very accurate results. For the power-law profile, it is possible to calculate the pulse broadening due to the fact that different rays take different amounts of time in traversing a certain length of the fiber; details can be found in many textbooks [see, e.g., Ghatak and Thyagarajan (1998)]. Here we will derive the results for a step-index fiber and just give the results for a parabolic index fiber.

Ray Dispersion in Step-Index Fibers

We first consider ray paths in a step-index fiber as shown in Fig. 6.4. As can be seen, rays making larger angles with the axis (those shown as dotted rays) have to traverse a longer optical path length and therefore take longer to reach the output end. We now derive an expression for the intermodal dispersion for a step-index fiber. Referring to Fig. 6.4, we note that for a ray making an angle θ with the axis, the distance AB is traversed in time

$$t_{AB} = \frac{AC + CB}{c/n_1} = \frac{AB/\cos\theta}{c/n_1}$$

or

$$t_{AB} = \frac{n_1 AB}{c \cos\theta}$$

where c/n_1 represents the speed of light in a medium of refractive index n_1, c being the speed of light in free space. Since the ray path will repeat itself, the time taken by a ray to traverse a length L of the fiber would be

$$t_L = \frac{n_1 L}{c \cos\theta} \tag{6.6}$$

This expression shows that the time taken by a ray is a function of the angle θ made by the ray with the z axis, which leads to pulse dispersion. If we assume that all rays

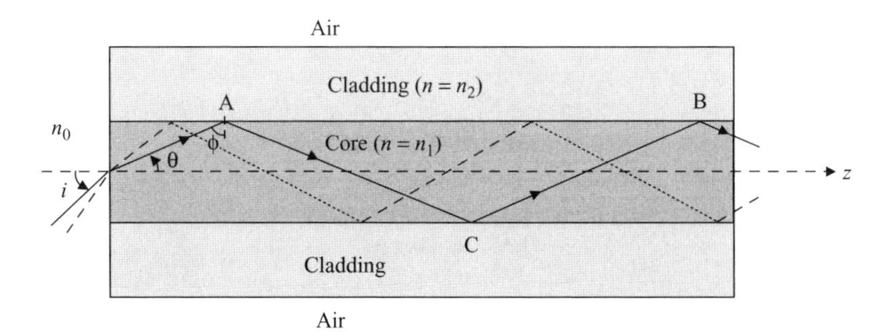

FIGURE 6.4 Ray paths in a step-index fiber.

lying between $\theta = 0$ and $\theta = \theta_c = \cos^{-1}(n_2/n_1)$ are present, the time taken by these extreme rays for a fiber of length L would be given by

$$t_{\min} = \frac{n_1 L}{c} \quad \text{corresponding to } \theta = 0 \tag{6.7}$$

$$t_{\max} = \frac{n_1^2 L}{c n_2} \quad \text{corresponding to } \theta = \theta_c = \cos^{-1}(n_2/n_1) \tag{6.8}$$

Hence, if all the input rays were excited simultaneously, at the output end the rays would occupy a time interval of duration

$$\Delta\tau_i = t_{\max} - t_{\min} = \frac{n_1 L}{c}\left(\frac{n_1}{n_2} - 1\right)$$

or

$$\Delta\tau_i \cong \frac{n_1 L}{c}\Delta \approx \frac{L}{2n_1 c}(\text{NA})^2 \tag{6.9}$$

where Δ is as defined earlier [see Eqs. (6.2) and (6.4)]. The quantity $\Delta\tau_i$ represents the pulse dispersion due to different rays taking different times in propagating through the fiber, which in wave optics is nothing but the intermodal dispersion—hence the subscript i. Note that the pulse dispersion is proportional to the square of NA. Thus, to have a smaller dispersion, one must have a smaller NA, which of course decreases the acceptance angle and hence the light-gathering power. Now, if at the input end of the fiber, we have a pulse of width τ_1, then after propagating through a length L of the fiber, the pulse would have a width τ_2 given approximately by

$$\tau_2^2 = \tau_1^2 + \Delta\tau_i^2 \tag{6.10}$$

If the input pulse width is 10 ns and the intermodal dispersion is 50 ns, the output pulse width would be approximately 51 ns. Thus, due to different velocities, the pulse broadens as it propagates through the fiber. Hence, even though two pulses may be well resolved at the input end, because of the broadening of the pulses they may not be so at the output end (Fig. 6.5).

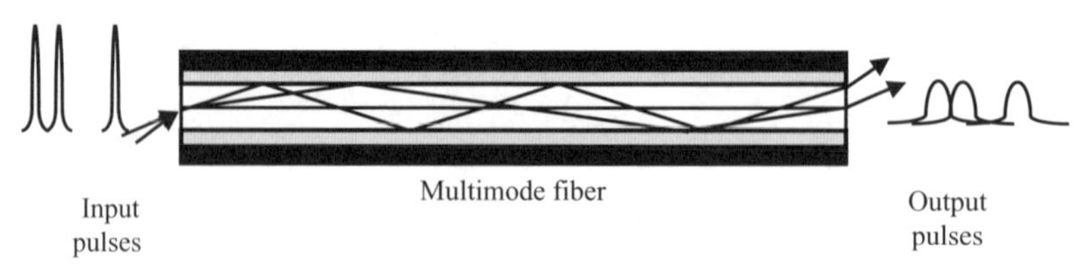

FIGURE 6.5 An incident pulse excites all rays, which then take different times to propagate, resulting in pulse broadening.

Example 6.1 For a typical (multimoded) step-index fiber, if we assume that $n_1 = 1.5$, $\Delta = 0.01$, and $L = 1$ km, we would get

$$\Delta\tau_i = \frac{(1.5)(1000)}{3 \times 10^8}(0.01) = 50\,\text{ns/km}$$

that is, a pulse will be broadened by 50 ns after traversing through the fiber of length 1 km. Thus, two pulses separated by, say, 500 ns at the input end would be quite resolvable at the end of 1 km of the fiber. However, if consecutive pulses are separated by, say, 10 ns at the input end, they would be absolutely unresolvable at the output end. Hence, in a 1-Mb/s fiber optic system, where we have one pulse every 10^{-6} s, a 50-ns/km dispersion would require repeaters to be placed every 3 to 4 km. On the other hand, in a 1-Gb/s fiber optic communication system, which requires the transmission of one pulse every 10^{-9} s, a dispersion of 50 ns/km would result in intolerable broadening even within 50 m or so, which would be highly inefficient and uneconomical from a system point of view.

Where the output pulses are not resolvable, no information can be retrieved. Thus, the smaller the pulse dispersion, the greater will be the information-carrying capacity of the system. From the discussion in Example 6.1 it follows that for a very high information-carrying system, it is necessary to reduce the pulse dispersion. Two alternative solutions exist: One involves the use of nearly parabolic index fibers (discussed in the next section), and the other involves single-mode fibers (discussed in Chapter 7).

Parabolic-Index Fibers

In a step-index fiber such as that pictured in Fig. 6.4, the refractive index of the core has a constant value. By contrast, in a *parabolic-index fiber*, the refractive index in the core decreases continuously (in a quadratic fashion) from a maximum value at the center of the core to the core–cladding interface, beyond which it is constant. The refractive index variation is given by

$$n^2(r) = \begin{cases} n_1^2\left[1 - 2\Delta\left(\dfrac{r}{a}\right)^2\right] & 0 < r < a \quad \text{core} \\ n_2^2 = n_1^2(1 - 2\Delta) & r > a \quad \text{cladding} \end{cases} \tag{6.11}$$

with Δ as defined in Eq. (6.2). For a typical (multimode) parabolic-index silica fiber, $\Delta \approx 0.01$, $n_2 \approx 1.45$, and $a \approx 25$ μm. Figure 6.6 shows the refractive index profile of a commercially available multimode fiber from Corning (InfiniCor 300 fiber) with a core diameter of 62.5 μm and an NA of 0.275. The fiber has an attenuation of 3 dB/km at 850 nm and 0.7 dB/km at 1300 nm.

When light rays propagate through a medium whose refractive index changes continuously, light rays do not travel along straight lines but get curved. The phenomenon of mirage is a manifestation of this property. In the case of parabolic-index fibers, it

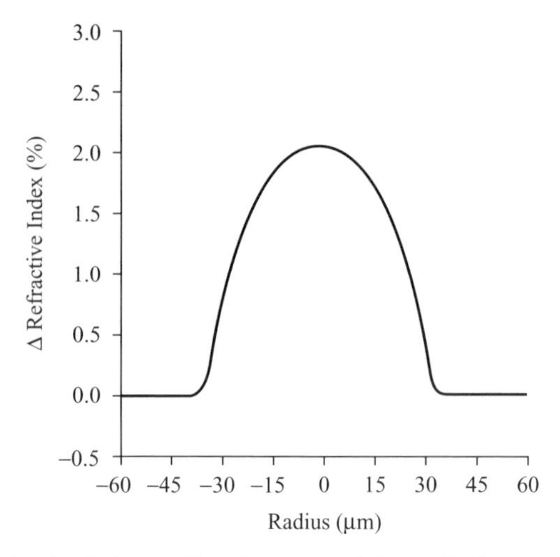

FIGURE 6.6 Refractive index profile of a commercially available graded-index multimode fiber from Corning.

can be shown that the ray paths are sinusoidal (Fig. 6.7). Note that many rays do not even propagate up to the core–cladding interface but turn back before reaching the interface.

Now, even though rays making larger angles with the axis traverse a larger path length, they do so now in a region of lower refractive index (and hence greater speed). The longer path length of these rays is compensated for almost completely by a greater average speed, such that all rays take approximately the same amount of time in traversing the fiber. The final result for intermodal dispersion is given by

$$\Delta\tau_i = \frac{n_2 L}{2c}\left(\frac{n_1 - n_2}{n_2}\right)^2 \approx \frac{n_2 L}{2c}\Delta^2 \approx \frac{L}{8cn_1^3}(\text{NA})^4 \tag{6.12}$$

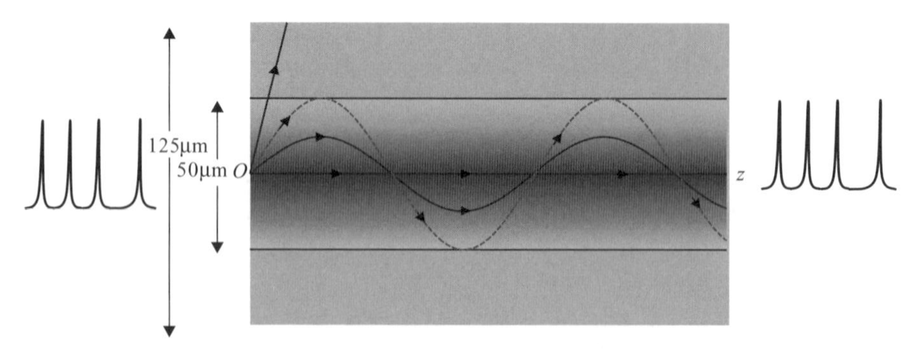

FIGURE 6.7 Ray paths in a parabolic-index fiber, leading to very limited ray dispersion.

Note that compared to a step-index fiber, in a parabolic index fiber the pulse dispersion is proportional to the fourth power of NA. For a typical (multimode parabolic index) fiber with $n_2 \approx 1.45$ and $\Delta \approx 0.01$ (NA ~ 0.2), we would get $\Delta\tau_i \approx 0.25$ ns/km. Referring to Eq. (6.9), we find that compared with the step-index fiber, pulse dispersion for a parabolic-index fiber is reduced by a factor of $4n_1^2/(\text{NA})^2$, which for the present example is about 200. It is for this reason that first- and second-generation optical communication systems used nearly parabolic refractive index fibers. Short-distance applications using plastic multimode fibers also use nearly parabolic refractive index profiles for enhanced bandwidths.

To further decrease the pulse dispersion, it is necessary to use single-mode fibers because there will be no intermodal dispersion. However, in all fiber optic systems we will have material dispersion, which is a characteristic of the material itself and not of the waveguide; we discuss this in Section 6.3.

6.3 MATERIAL DISPERSION

We mentioned in Chapter 2 that the refractive index of a medium depends on the wavelength of light. To recall, a beautiful natural effect of this is the formation of rainbows in the sky (Fig. 2.14). Similarly, when a narrow pencil of a beam of white light is incident on the face of a prism, since the refractive index of glass depends on wavelength, the angle of refraction will be different for different colors, and as shown in Fig. 2.13, the incident white light will *disperse* (or split) into its constituent colors. Optical fibers made of silica also have different refractive indices for different wavelengths. Since the time taken by a light signal to propagate through a given length of the fiber depends on the speed of propagation and hence on the refractive index, this would imply that different wavelengths would take different times to propagate through a given length of fiber.

The peak position of a temporal pulse travels with what is known as the *group velocity* (v_g), which is given by

$$v_g = \frac{c}{n_g} \tag{6.13}$$

where n_g is known as the *group refractive index* and is given by

$$n_g = n(\lambda_0) - \lambda_0 \frac{dn}{d\lambda_0} \tag{6.14}$$

and $n(\lambda_0)$ represents the refractive index, which depends on the wavelength. In Table 6.1 we have tabulated $n(\lambda_0)$ and $n_g(\lambda_0)$ for pure silica as a function of the free-space wavelength λ_0. In Fig. 6.8 we have plotted the wavelength variation of the group velocity v_g for pure silica; notice that the group velocity attains a maximum value at $\lambda_0 \approx 1.27$ μm. As we show later, this wavelength is of great significance in optical communication systems.

TABLE 6.1 Values of n, n_g, and D_m for Pure Silica

λ_0 (μm)	$n(\lambda_0)$	$\frac{dn}{d\lambda_0}\,\mu m^{-1}$	$n_g(\lambda_0)$	$\frac{d^2n}{d\lambda_0^2}\,\mu m^{-2}$	D_m (ps/nm·km)
0.70	1.45561	−0.02276	1.47154	0.0741	−172.9
0.75	1.45456	−0.01958	1.46924	0.0541	−135.3
0.80	1.45364	−0.01725159	1.46744	0.0400	−106.6
0.85	1.45282	−0.01552236	1.46601	0.0297	−84.2
0.90	1.45208	−0.01423535	1.46489	0.0221	−66.4
0.95	1.45139	−0.01327862	1.46401	0.0164	−51.9
1.00	1.45075	−0.01257282	1.46332	0.0120	−40.1
1.05	1.45013	−0.01206070	1.46279	0.0086	−30.1
1.10	1.44954	−0.01170022	1.46241	0.0059	−21.7
1.15	1.44896	−0.01146001	1.46214	0.0037	−14.5
1.20	1.44839	−0.01131637	1.46197	0.0020	−8.14
1.25	1.44783	−0.01125123	1.46189	0.00062	−2.58
1.30	1.44726	−0.01125037	1.46189	−0.00055	2.39
1.35	1.44670	−0.01130300	1.46196	−0.00153	6.87
1.40	1.44613	−0.01140040	1.46209	−0.00235	10.95
1.45	1.44556	−0.01153568	1.46229	−0.00305	14.72
1.50	1.44498	−0.01170333	1.46253	−0.00365	18.23
1.55	1.44439	−0.01189888	1.46283	−0.00416	21.52
1.60	1.44379	−0.01211873	1.46318	−0.00462	24.64

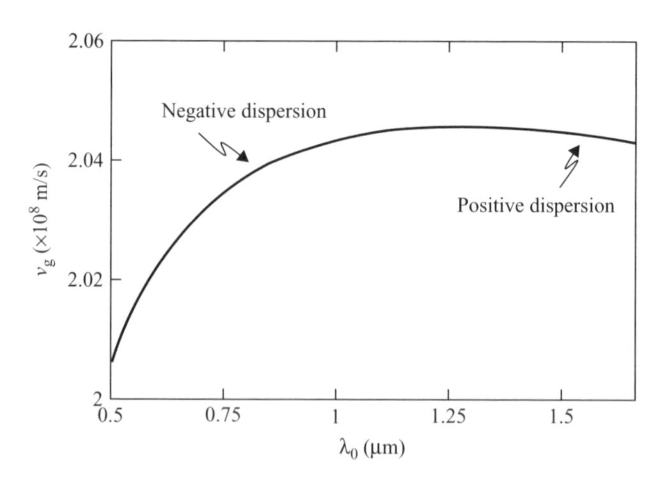

FIGURE 6.8 Wavelength dependence of group velocity for a silica fiber.

Example 6.2 For pure silica the refractive index variation in the wavelength domain 0.5 μm < λ_0 < 1.6 μm can be assumed to be given by the following approximate empirical formula:

$$n(\lambda_0) \approx C_0 - a\lambda_0^2 + \frac{a}{\lambda_0^2} \tag{6.15}$$

where $C_0 \approx 1.451$, $a \approx 0.003$, and λ_0 is measured in micrometers. Simple algebra shows that

$$n_g(\lambda_0) = C_0 + a\lambda_0^2 + \frac{3a}{\lambda_0^2} \tag{6.16}$$

Thus, at $\lambda_0 = 1\ \mu\text{m}$,

$$n(\lambda_0) \approx 1.451$$

and

$$n_g(\lambda_0) \approx 1.463$$

indicating that the difference between group and phase velocities is about 0.8%. More accurate values of $n(\lambda_0)$ and $n_g(\lambda_0)$ are given in Table 6.1.

Using Table 6.1, we find that in pure silica, for $\lambda_0 = 0.8\ \mu\text{m}$,

$$v_g = \frac{c}{n_g} = 2.0444 \times 10^8 \text{m/s}$$

and for $\lambda_0 = 0.85\ \mu\text{m}$,

$$v_g = \frac{c}{n_g} = 2.0464 \times 10^8 \text{m/s}$$

implying that higher-wavelength components travel faster, referred to as *normal group velocity dispersion*. Similarly, the group velocities at 1.3 and 1.35 μm are 2.0521×10^8 and 2.0520×10^8 m/s, respectively, indicating that higher wavelengths travel more slowly then the lower wavelengths, referred to as *anomalous group velocity dispersion*. The switchover of velocities takes place at the zero-dispersion wavelength, which is close to 1.27 μm for pure silica.

Now every source of light would have a certain wavelength spread, which is usually referred to as the *spectral width* of the source. Thus, a white light source (e.g., one coming from the sun) would have a spectral width of about 300 nm; on the other hand, a light-emitting diode (LED) would have a spectral width of about 25 nm, and a typical laser diode (LD) operating around 1.3 μm would have a spectral width of about 2 nm; this spectral width is usually denoted by $\Delta\lambda_0$. Since each wavelength component (of a pulse) will travel with a slightly different group velocity, it will, in general, result in broadening of the pulse.

To calculate the broadening, we note that the time taken by a pulse to traverse a length L of the dispersive medium is given by

$$\tau = \frac{L}{v_g} = \frac{L}{c}\left[n(\lambda_0) - \lambda_0 \frac{dn}{d\lambda_0}\right] \tag{6.17}$$

Since the right-hand side depends on λ_0, Eq. (6.17) implies that different wavelengths will travel with different group velocities in propagating through a certain length of the dispersive medium. Thus, the pulse broadening will be given by

$$\Delta\tau_m = \frac{d\tau}{d\lambda_0}\Delta\lambda_0 = -\frac{L}{c}\frac{\Delta\lambda_0}{\lambda_0}\left(\lambda_0^2\frac{d^2n}{d\lambda_0^2}\right) \tag{6.18}$$

where $\Delta\lambda_0$ is the spectral width of the source. The quantity $\Delta\tau_m$ is usually referred as *material dispersion* because it is due to the material properties of the medium—hence the subscript m. In Eq. (6.18), the quantity inside parentheses is dimensionless. Indeed, after propagating through a length L of the dispersive medium, a pulse of temporal width τ_0 will get broadened to τ_f, where

$$\tau_f^2 \approx \tau_0^2 + (\Delta\tau_m)^2 \tag{6.19}$$

From Eq. (6.18) we see that the broadening of the pulse is proportional to the length L traversed in the medium and also to the spectral width of the source, $\Delta\lambda_0$.

We define the material dispersion coefficient D_m (in ps/km·nm) as the pulse broadening in picoseconds caused by material dispersion in propagating through 1 km of fiber when the source spectral width is 1 nm. Thus, using $L = 1000$ m and $\Delta\lambda_0 = 1$ nm in Eq. (6.18), we obtain

$$D_m = \frac{\Delta\tau_m}{L\,\Delta\lambda_0} \approx -\frac{1}{3\lambda_0}\left(\lambda_0^2\frac{d^2n}{d\lambda_0^2}\right) \times 10^4\ \text{ps/km}\cdot\text{nm} \tag{6.20}$$

where λ_0 is measured in micrometers and we have assumed that $c \approx 3 \times 10^8$ m/s. A medium is said to be characterized by *positive dispersion* when D_m is positive, and is said to be characterized by *negative dispersion* when D_m is negative. Positive dispersion implies that longer wavelengths take longer to propagate, and hence it corresponds to anomalous group velocity dispersion. Similarly, negative dispersion implies normal group velocity dispersion. Note that whether the dispersion coefficient is positive or negative, the pulse will undergo broadening.

We mention here that the spectral width of a pulse is usually due to the intrinsic spectral width of the source, which for a typical LED is about 25 nm and for a commercially available laser diode is about 1 to 2 nm. On the other hand, for a nearly monochromatic source, the intrinsic spectral width could be extremely small, and the actual spectral width of a pulse is determined from its finite duration (such a pulse is often referred to as a *Fourier-transform limited pulse*). Thus, a 20-ps pulse will have a spectral width of

$$\Delta\nu = \frac{\Delta w}{2\pi} = \frac{1}{2\pi\tau_0} \simeq 8 \times 10^9\ \text{Hz}$$

implying that

$$\Delta\lambda_0 \approx \frac{\lambda_0^2\,\Delta\nu}{c} \approx 0.06\,\text{nm}$$

Hence, even pulses generated out of extremely monochromatic sources would have a finite spectral width, due to the finite duration of the pulse.

Example 6.3 A first-generation optical communication system used LEDs with $\lambda_0 \approx 0.85\ \mu\text{m}$ and $\Delta\lambda_0 \approx 25$ nm. At $\lambda_0 \approx 0.85\ \mu\text{m}$,

$$\frac{d^2n}{d\lambda_0^2} \approx 0.030\ \mu\text{m}^{-2}$$

giving

$$D_m \approx -85\ \text{ps/km}\cdot\text{nm}$$

the negative sign indicating that higher wavelengths travel faster than lower wavelengths. Thus, for $\Delta\lambda_0 \approx 25$ nm, the actual broadening of the pulse will be $\Delta\,\tau_{\text{m}} \approx$ 2.1 ns/km, implying that a very narrow pulse will broaden by 2.1 ns after traversing through 1 km of the silica fiber.

Example 6.4 Fourth-generation optical communication systems had laser diodes with $\lambda_0 = 1.55\ \mu\text{m}$ and $\Delta\lambda_0 \approx 2$ nm. Now at $\lambda_0 \approx 1.55\ \mu\text{m}$,

$$\frac{d^2n}{d\lambda_0^2} \approx 0.0042\ \mu\text{m}^{-2}$$

giving

$$D_m \approx +21.7\ \text{ps/km}\cdot\text{nm}$$

the positive sign indicating that higher wavelengths travel more slowly than do lower wavelengths. (Notice from Table 6.1 that for $\lambda_0 \geq 1.27\ \mu\text{m}$, n_g increases with λ_0). Thus, for $\Delta\lambda_0 \approx 2$ nm, the actual broadening of the pulse will be

$$\Delta\tau_m \approx 43\ \text{ps/km}$$

implying that a very narrow pulse will broaden by 43 ps after traversing through 1 km of the silica fiber.

We see that for pure silica,

$$\frac{d^2n}{d\lambda_0^2} \approx 0$$

around $\lambda_0 \approx 1.27\ \mu\text{m}$. Indeed, the wavelength $\lambda_0 \approx 1270$ nm is usually referred to as the *zero material dispersion wavelength* and it is in view of low material dispersion, second- and third-generation optical communication systems operated around $\lambda_0 \simeq$ 1300 nm.

6.4 LASER-OPTIMIZED MULTIMODE OPTICAL FIBERS

There exists a large network of multimode fibers installed for short-distance communications such as intrabuilding cabling, and more than 75% of these are within 300-m distances. These conventional multimode fibers were designed to operate with light-emitting diodes which fill the entire numerical aperture of the multimode fiber, thus exciting all the modes. As we have seen earlier, this leads to intermodal dispersion, and thus such a fiber does not support high-speed communication at, say, 10-Gb/s rates to distances of 300 m. To generate an optical pulse stream at a data rate of 10 Gb/s, we use laser diodes, and in the case of multimode fibers, the lasers used are vertical cavity surface emitting lasers (VCSEL; see Chapter 8). The small spectral width of the laser also contributes to a decrease in material dispersion and hence an increase in capacity. Unlike LEDs, laser diode emissions are much more directional and hence do not excite all possible rays or modes of the fiber. Figure 6.9 shows the difference in the launch between an LED and a laser diode, pointing clearly to a very much reduced region of coupling. In fact, since the laser spot size is much smaller than the core diameter of multimode fibers (typically, 50 and 62.5 μm), excitation of the modes would change with the positioning of the laser diode in front of the multimode fiber.

As we discussed earlier, the intermodal dispersion arises due to the different propagating times of the various rays. The propagating times of rays depend on the refractive index profile of the multimode fiber. Conventional methods of measuring the refractive index profile to estimate the bandwidth of the multimode fiber work well for LED inputs, which excite all modes. In contrast, a laser source leads to excitation of only certain groups of rays, and even small variations in the refractive index profile can give large differences in the ray dispersion value and hence the bandwidth. A better way to characterize the fiber for its bandwith capability is the technique of measuring the differential mode delay (DMD). In this technique the multimode fiber is excited

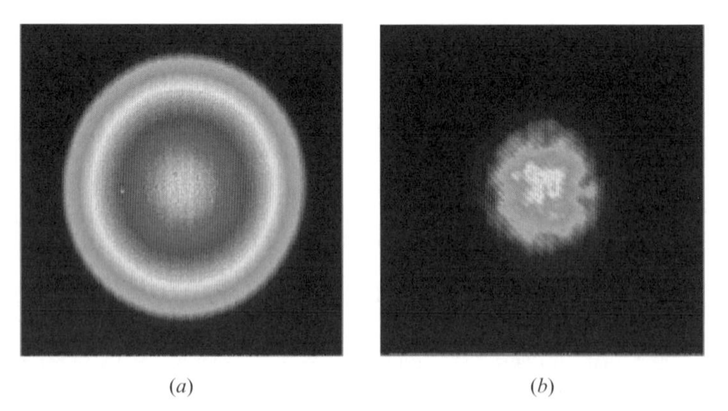

(*a*) (*b*)

FIGURE 6.9 Excitation of a multimode fiber using an LED (*a*) and a laser diode (*b*). (Adapted from Pondillo, 2001.)

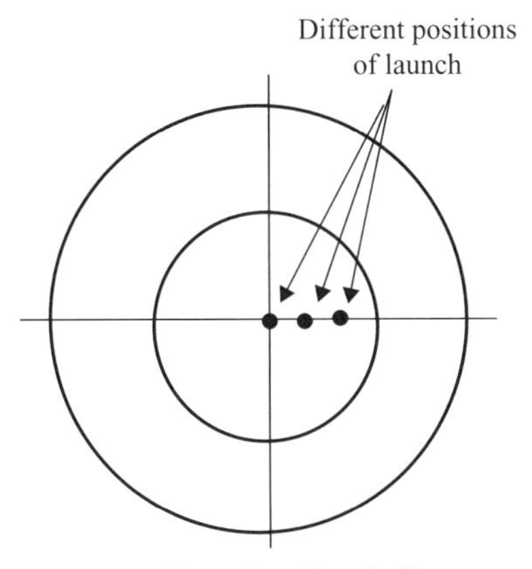

FIGURE 6.10 Differential mode delay is measured by exciting the multimode fiber with a small spot and measuring the propagation delay as a function of the position of the laser spot.

by a short pulse, which covers a very small area (of radius of about 5 to 7 μm) at the entrance face of the fiber, thus exciting only a specific group of rays (Fig. 6.10). As the illuminating spot is moved across the cross section of the fiber, the laser spot would excite different ray groups. Measuring the time of arrival as a function of the position of the illuminating spot gives us a measure of the differential mode delay of the fiber (Fig. 6.11). For an optimized fiber, this differential mode delay should be very small.

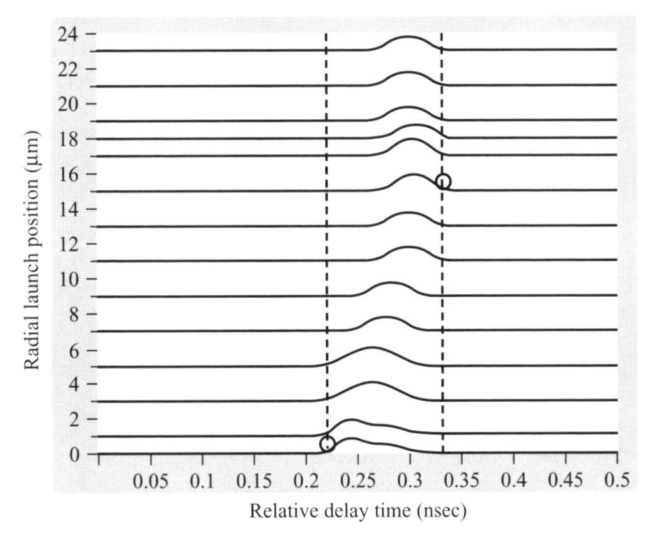

FIGURE 6.11 Typical differential mode delay measurement. The smallest time corresponds to launch at $r = 0$ μm and the largest time to launch at $r = 15$ μm. (After Kritler, 2003.)

Thus, by using laser-optimized multimode fibers exhibiting smaller dispersion, it is possible to achieve bit rates of 10 Gb/s over a distance of 300 m. Demonstrations of 40-Gb/s systems over a 400-m propagation distance using multimode fibers at 1300 nm have also been reported (Matthijsse et al., 2006).

6.5 DISPERSION AND MAXIMUM BIT RATE

We mention here briefly that in a digital communication system employing light pulses, pulse broadening would result in an overlap of pulses, resulting in a loss of resolution and leading to errors in detection. Thus, pulse broadening is one of the mechanisms (other than attenuation) that limits the distance between two repeaters in a fiber optic link. It is obvious that the larger the pulse broadening, the smaller will be the number of pulses per second that can be sent down a link. Different criteria based on slightly different considerations are used to estimate the maximum permissible bit rate ($B_{\max}$) for a given pulse dispersion. However, it is always of the order of $1/\Delta\tau$. In one type of coding used extensively [known as NRZ (non return to zero); see Chapter 3], we have

$$B_{\max} \approx \frac{0.7}{\Delta\tau} \tag{6.21}$$

This formula takes into account (approximately) only the limitation imposed by the pulse dispersion in the fiber. In an actual link the source and detector characteristics would also be taken into account when estimating the maximum bit rate. It should also be pointed out that in a fiber, the pulse dispersion is caused, in general, by intermodal dispersion, material dispersion, and waveguide dispersion. However, waveguide dispersion is important only in single-mode fibers and may be neglected in carrying out analysis for multimode fibers. Thus (considering multimode fibers), if $\Delta\tau_i$ and $\Delta\tau_m$ are the dispersion due to intermodal and material dispersions, respectively, the total dispersion is given

$$\Delta\tau = \sqrt{(\Delta\tau_i)^2 + (\Delta\tau_m)^2} \tag{6.22}$$

Example 6.5 We consider a step-index multimode fiber with $n_1 = 1.46$, $\Delta = 0.01$ operating at 850 nm. For such a fiber, $\Delta\tau_i \approx 49$ ns/km; further, if the source is an LED with $\Delta\lambda = 20$ nm, then $\Delta\tau_m \approx 1.7$ ns/km, giving $\Delta\tau \approx 49$ ns/km. This gives a maximum bit rate of about

$$B_{\max} \approx \frac{0.7}{\Delta\tau} = \frac{0.7}{49 \times 10^{-9}}\,\text{b}\cdot\text{km/s} \approx 14\,\text{Mb}\cdot\text{km/s}$$

Thus, a 10-km link can support at most only 1.4 Mb/s.

Example 6.6 We next consider a parabolic-index multimode fiber with $n_1 = 1.46$, $\Delta = 0.01$ operating at 850 nm with an LED of spectral width 20 nm. For such a fiber, the intermodal dispersion $\Delta\tau_i \approx 0.24$ ns/km; the material dispersion is again

1.7 ns/km. Thus, in this case the dominant mechanism is material dispersion rather than intermodal dispersion. The total dispersion is $\Delta\tau \approx 1.72$ ns/km. This gives a maximum bit rate of about

$$B_{\max} \approx \frac{0.7}{1.72 \times 10^{-9}} \text{ b·km/s} \approx 400 \text{ Mb} \cdot \text{km/s}$$

giving a maximum permissible bit rate of 20 Mb/s for a 20-km link.

If we now shift the wavelength of operation to 1300 nm and use the parabolic-index fiber of Example 6.6, we see that the intermodal dispersion remains 0.24 ns/km, while the material dispersion (for an LED of $\Delta\lambda_0 = 20$ nm) becomes 0.05 ns/km. The material dispersion is now negligible compared to the intermodal dispersion. Thus, the total dispersion and maximum bit rate are given by $\Delta\tau = \sqrt{0.24^2 + 0.05^2} = 0.25$ ns/km $\Longrightarrow B_{\max} = 2.8$ Gb·km/s. Thus, in principle, bit rates of greater than 5 Gb/s over distances of about 500 m are possible with multimode fibers.

We should reiterate that in the examples discussed above, the maximum bit rate has been estimated by considering the fiber only. In an actual link, the temporal response of the source and detector must also be taken into account.

CHAPTER 7

Pulse Dispersion in Single-Mode Optical Fibers

7.1 INTRODUCTION

As discussed in Chapter 6, pulse dispersion arises due primarily to:

- *Ray dispersion*, which is caused by the fact that different rays take different times to propagate (along different paths) from the input end to the output end of the fiber; this is also known as *intermodal dispersion*.
- *Material dispersion*, which is caused by the fact that the refractive index of the material of the fiber depends on the wavelength and that the source always has a finite spectral width.

To decrease pulse dispersion further, it is necessary to use single-mode fibers. As mentioned in Chapter 4, we can imagine a single-mode fiber allowing propagation of only one light ray path, corresponding to a single mode, and therefore we would not have any ray (or intermodal) dispersion. However, in single-mode fibers, in addition to material dispersion, another dispersion mechanism, referred to as *waveguide dispersion*, becomes important; waveguide dispersion depends on the transverse refractive index profile. In the next section we discuss how by changing the transverse refractive index profile, we can change the wavelength dependence of the total dispersion. Later, we discuss dispersion-compensating fibers, which are now used extensively in optical fiber communication systems. Further, when we refer to high bit rates, it is also important to consider the phenomenon of *polarization mode dispersion,* which can limit the capacity of single-mode fiber optic systems; this is discussed briefly at the end of the chapter.

7.2 WAVEGUIDE DISPERSION

In Chapter 6, we discussed material dispersion, which results from the dependence on wavelength of the refractive index of the fiber material. Even if we assume the core

Fiber Optic Essentials, By K. Thyagarajan and Ajoy Ghatak

and cladding refractive indices to be independent of wavelength, the group velocity of each mode would still depend on the wavelength (due to the geometry of the fiber). This leads to what is known as *waveguide dispersion*, which together with material dispersion is referred to as *intramodal dispersion* (broadening within a mode); ray dispersion is referred to as *intermodal dispersion* (dispersion between different modes). In multimode fibers, intermodal dispersion dominates over intramodal dispersion and the latter can safely be neglected. Like material dispersion, waveguide dispersion is proportional to the spectral width of the source.

Physically, waveguide dispersion arises due to the fact that the spot size of the mode depends on the wavelength (see Examples 4.4 and 4.5), and therefore the fractional power within the core and the cladding changes with wavelength, which in turn leads to dependence of the group velocity on the wavelength. Thus, by modifying the transverse refractive index profile of a fiber, we can change the wavelength dependence of the spot size and hence modify the waveguide dispersion.

The waveguide dispersion D_w in a step-index single-mode fiber can be approximated by the expression (see, e.g., Ghatak and Thyagarajan, 1998)

$$D_w = -\frac{n_2 \Delta}{3\lambda_0} \times 10^7 \left[0.080 + 0.549(2.834 - V)^2\right] \qquad \text{ps/km} \cdot \text{nm} \tag{7.1}$$

where λ_0 is the operating wavelength measured in nanometers and other parameters are the same as defined earlier. D_w measured in units of ps/km·nm gives the pulse dispersion in picoseconds suffered by a pulse in propagating through 1 km of fiber when the source has a spectral width of 1 nm.

As before, the negative sign indicates that longer wavelengths travel faster than do shorter wavelengths. The total dispersion is given approximately by the sum of material and waveguide dispersions:

$$D_{\text{tot}} = D_m + D_w \tag{7.2}$$

Since material dispersion passes through a zero value around 1270 nm and changes sign, it is possible to cancel material dispersion with waveguide dispersion (or vice versa), resulting in zero total dispersion. The wavelength where this happens, the *zero-dispersion wavelength*, is a very important quantity specifying single-mode optical fibers.

Let us next consider the two single-mode fibers discussed in Chapter 4.

Example 7.1 We first consider the fiber discussed in Example 4.2, for which $n_2 = 1.447$, core radius $a = 4.2$ μm, with $\Delta = 0.003$, so that the cladding index is about 0.3% less than the core (the G.652 fiber); thus, $V = 2958/\lambda_0$, where λ_0 is measured in nanometers. Substituting these values in Eq. (7.2), we get

$$D_w = -\frac{1.447 \times 10^4}{\lambda_0} \left[0.080 + 0.549\left(2.834 - \frac{2958}{\lambda_0}\right)^2\right] \text{ ps/km·nm}$$

Elementary calculations show that at $\lambda_0 \approx 1300$ nm, $D_w \approx -2.8$ ps/km·nm, and since at this wavelength $D_m \approx +2.4$ ps/km·nm, the total dispersion would be almost zero close to $\lambda_0 \approx 1300$ nm, which is the zero-dispersion wavelength of this fiber. For this fiber, at $\lambda_0 \approx 1550$ nm, $D_w \approx -5.1$ ps/km·nm and $D_m \approx +21.5$ ps/km·nm, giving a total dispersion of about 16.4 ps/km·nm; these numbers are just to give approximate variations of various quantities.

Example 7.2 We next consider the fiber discussed in Example 4.3, for which $n_2 = 1.444$, $\Delta = 0.0075$, and $a = 2.3\ \mu$m, so that $V = 2556/\lambda_0$, where, once again, λ_0 is measured in nanometers. Substituting in Eq. (4.1), we get

$$D_w = -\frac{3.61 \times 10^4}{\lambda_0}\left[0.080 + 0.549\left(2.834 - \frac{2556}{\lambda_0}\right)^2\right] \text{ps/km·nm}$$

Thus, at $\lambda_0 \approx 1550$ nm,

$$D_w = -20 \text{ ps/km·nm}$$

On the other hand, the material dispersion at this wavelength is given by (see Table 6.1)

$$D_m = +21.52 \text{ ps/km·nm}$$

Thus, this fiber would have a very small dispersion around 1550 nm.

We therefore see that it is possible that the material and waveguide dispersions will have opposite signs, thus almost canceling each other, leading to nearly zero dispersion. Physically, because of waveguide dispersion, longer wavelengths travel more slowly than do shorter wavelengths, and because of material dispersion, longer wavelengths travel faster than do shorter wavelengths—and the two effects can cancel, resulting in zero total dispersion.

Thus, by modifying the transverse refractive index profile of a fiber, we can modify the waveguide dispersion property of the fiber. In Fig. 7.1 we have shown the wavelength dependence of the material dispersion (labeled as D_m) and of the waveguide dispersion (labeled as D_w), for two single-mode fibers, one corresponding to a core of radius 4.2 μm with a cladding index about 0.3% less than the core (the G.652 fiber), and the other corresponding to a core of radius 2.3 μm with a cladding index about 0.75% less than the core; the variation in D_m is calculated using Table 6.1. Since material dispersion is always present, the total dispersion would be given by the sum of the waveguide and material dispersions. Figure 7.2 shows the total dispersion (labeled as $D_{tot} = D_m + D_w$) for two different types of fibers. For the G.652 fiber, it passes through zero when the wavelength is about 1310 nm. For the small-core fiber the total dispersion now passes through zero when the wavelength is about 1550 nm. Hence, it is possible to cancel material dispersion by waveguide

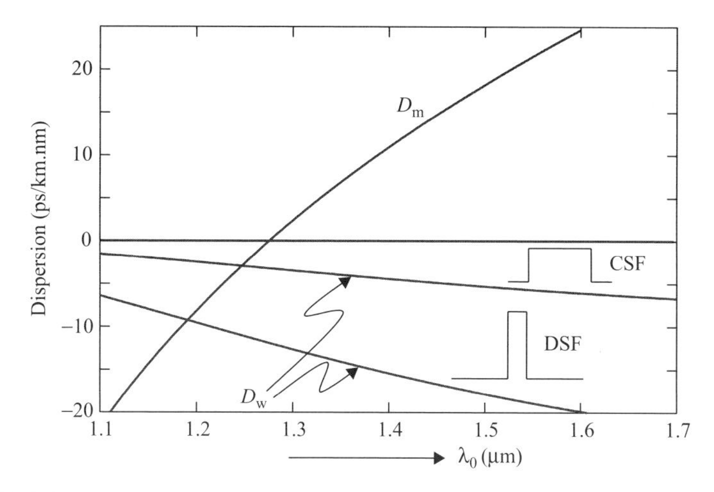

FIGURE 7.1 Wavelength dependence of material dispersion (labeled D_m) for a silica fiber. The other two curves represent the waveguide dispersion (labeled D_w) for a typical CSF (G.652) single-mode fiber and for a typical dispersion-shifted single-mode fiber.

dispersion (or vice versa) and achieve zero dispersion at a chosen wavelength. Thus, *by tailoring the transverse refractive index profile, we can change the wavelength dependence of waveguide dispersion and hence of total dispersion.*

Fibers with zero total dispersion at about 1300 nm are known as *conventional single-mode fibers* (CSFs; referred to according to telecommunication standards as

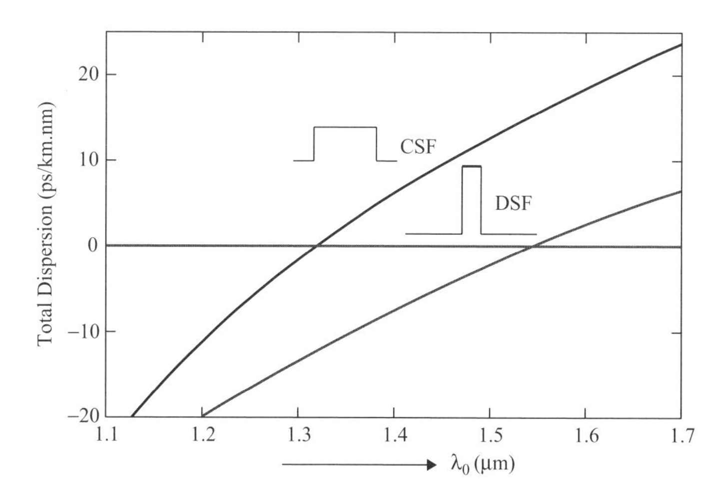

FIGURE 7.2 For a typical CSF (G.652) single-mode fiber, the total dispersion passes through zero around $\lambda_0 \approx 1300$ nm, which is known as a *zero total dispersion wavelength*. For a typical dispersion-shifted fiber (DSF or G.653 fiber), the total dispersion passes through zero around $\lambda_0 \approx 1550$ nm.

G.652 fibers). On the other hand, fibers with zero total dispersion at about 1550 nm (where the loss is lowest) are known as *dispersion-shifted fibers* (DSFs; referred to as G.653 fibers). Apart from these, the nonzero-dispersion-shifted fiber (NZ-DSF) is designed to have a nonzero finite dispersion at 1550 nm. This is to ensure that there is no nonlinear mixing (also referred to as *four-wave mixing*; see Chapter 13) between various wavelength channels. Such fibers are referred to as G.655 fibers.

Apart from dispersion coefficient D, a single-mode fiber is also specified by the slope of dispersion versus wavelength at the zero-dispersion wavelength. This is specified by the dispersion slope, which is measured in units of ps/km·nm^2 and usually denoted by S.

The wavelength dependence of dispersion of a single-mode fiber is usually described by the equation

$$D(\lambda) = \frac{S_0}{4}\left(\lambda - \frac{\lambda_z^4}{\lambda^3}\right) \tag{7.3}$$

where λ_z is the zero-dispersion wavelength and λ represents the operating wavelength. Equation (7.3) describes the dispersion of the fiber in the wavelength range 1200 to 1625 nm. For the G.652 fiber, the zero-dispersion wavelength λ_z lies between 1302 and 1322 nm, and the dispersion slope $S_0 < 0.092$ ps/km·nm^2.

Table 7.1 gives the values of dispersion and the dispersion slope of standard single-mode fibers operating at 1550 nm. Figure 7.3 shows the dispersion characteristics of some standard single-mode fibers that are available commercially.

Figure 7.4 shows the evolution of the pulse shape and pulse spectrum with distance of propagation in a standard single-mode fiber. The simulations have been carried out by neglecting attenuation in the fiber. Note that the pulse disperses in the time domain while its frequency spectrum remains the same. Dispersion causes various frequencies to propagate at different velocities, and all the frequencies that are input into the optical fiber also emerge from it.

It appears that when an optical fiber is operated at the zero-dispersion wavelength, the pulses will experience no dispersion. In fact, zero dispersion signifies only that the dispersion is approximately zero, since higher-order effects can still distort the pulse. Such effects become important for high-bit-rate systems operating very close to the zero-dispersion wavelength.

TABLE 7.1 Values of Dispersion and Dispersion Slope for Some Standard Fibers at 1550 nm

Fiber Type	D (ps/km·nm)	S (ps/km·nm^2)
Standard SMF (G.652)	17	0.058
LEAF (Corning)	4.2	0.085
Truewave-reduced slope (OFS)	4.5	0.045
TeraLight (Alcatel)	8.0	0.057

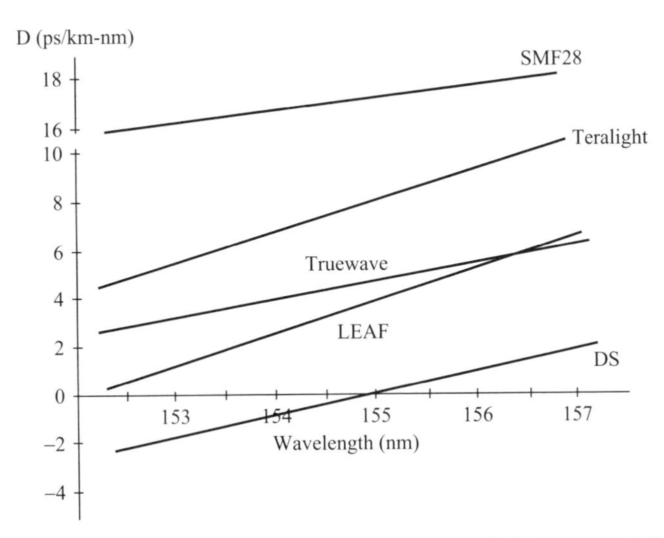

FIGURE 7.3 Various types of single-mode fibers that are available commercially. The fibers include the standard single-mode fiber, dispersion-shifted fiber and nonzero dispersion-shifted fiber.

Figure 7.5 shows the eye pattern of a non-return-to-zero (NRZ) pulse sequence at a bit rate of 10 Gb/s operating at a wavelength of 1551.7 nm after propagation through 500 and 8500 km of fiber. (Eye pattern is introduced in Chapter 8.) The experiment is performed in a recirculating loop consisting of 100 km of dispersion-shifted fiber with a dispersion coefficient $D = -1.2$ ps/km·nm and 7 km of standard fiber with

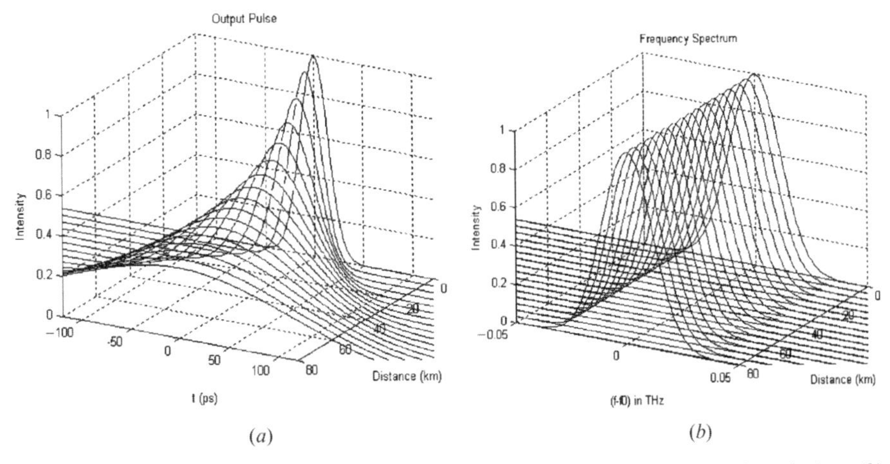

FIGURE 7.4 (*a*) Evolution of the pulse in the time domain showing pulse broadening; (*b*) the optical spectrum of the same pulse remains constant as it propagates. Dispersive effects do not change the spectrum.

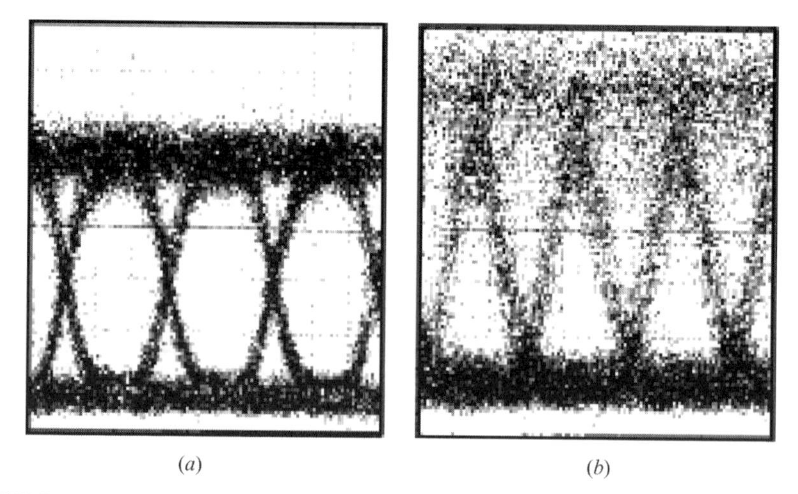

FIGURE 7.5 Eye pattern corresponding to an NRZ pulse train at 10 Gb/s after propagating through (*a*) 500 and (*b*) 8500 km of optical fiber. (Adapted from Jacob et al., 1997. Copyright © 1997 IEEE.)

$D = +16.5$ ps/km·nm at 1550 nm. The path-averaged dispersion is -0.02 ps/km·nm. The loop also contains erbium-doped fiber amplifiers for amplification, and the average power within the loop is about -4 dBm.

In Chapter 8 we discuss the effect of dispersion on the maximum bit rate that is possible to achieve using single-mode fibers.

7.3 DISPERSION-COMPENSATING FIBERS

When the possibility of achieving zero dispersion in the 1310-nm wavelength window was discovered, optical communication systems moved into operating at 1310 nm. Single-mode optical G.652 fibers, which have a zero-dispersion wavelength in the 1310-nm window, became the fiber of choice, and millions of kilometers of this fiber were laid throughout the world both under the ground and under the sea. Although these fibers had very little dispersion, the repeater spacing was now limited by attenuation due to an attenuation figure of about 1 dB/km at a wavelength of 1310 nm.

Erbium-doped fiber amplifiers (see Chapter 9), which can amplify multiple optical signals in the 1550-nm window simultaneously, became available in the early 1990s. It was also known that the single-mode fiber had the lowest loss in the 1550-nm window (see Chapter 5). Hence, an obvious shift of wavelength of operation to the 1550-nm wavelength window took place in the early 1990s. However, as discussed earlier, G.652 fiber has a dispersion of about $+17$ ps/km·nm in this wavelength window. Hence, operating this fiber in the 1550-nm window would, of course, overcome the problem of attenuation but would have the problem of dispersion. Although fibers

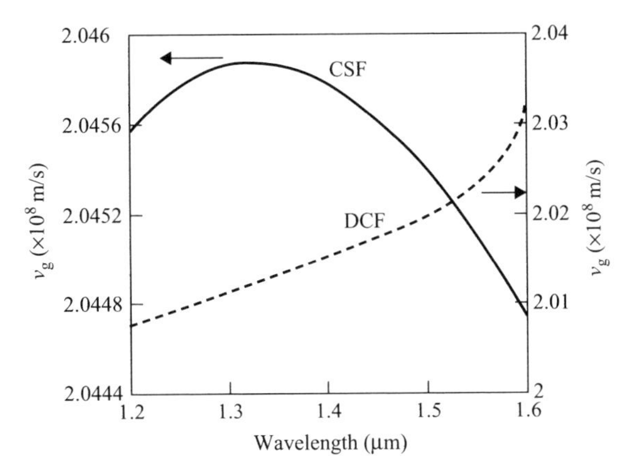

FIGURE 7.6 Wavelength variation of group velocity for a typical dispersion-compensating fiber and a typical conventional (G. 652) single-mode fiber.

with zero dispersion in the 1550-nm window became available, the cost of replacing the existing millions of kilometers of fiber would be exorbitant.

A more interesting solution was to compensate for the accumulated dispersion optically by passing the (dispersed) pulses through specially designed fibers, referred to as *dispersion-compensating fibers*, characterized by a large negative total dispersion.

To understand this phenomenon, we have plotted in Fig. 7.6 (as a solid curve) a typical variation of the group velocity v_g with wavelength for a conventional single-mode (G.652) fiber with zero dispersion around 1300-nm wavelength. As shown, v_g attains a maximum value at the zero-dispersion wavelength and on either side decreases monotonically with wavelength. Thus, if the central wavelength of the pulse is around 1550 nm, the red components of the pulse (i.e., longer wavelengths) will travel more slowly than the blue components (i.e., smaller wavelengths) of the pulse. Because of this, the pulse will be broadened. Now, after propagating through a G.652 fiber for a certain length L_1, the pulse is allowed to propagate through a length L_2 of a dispersion-compensating fiber (DCF), which is designed such that the group velocity v_g varies approximately as shown by the dashed curve in Fig. 7.6. The red components (i.e., longer wavelengths) will now travel faster than the blue components, and the pulse will tend to reshape itself into its original form. Indeed, if the lengths of the two fibers (L_1 and L_2), with dispersion coefficients D_1 and D_2, respectively, are such that

$$D_1 L_1 + D_2 L_2 = 0 \tag{7.4}$$

the pulse emanating from the second fiber will be almost identical to the pulse entering the first fiber, as shown in Fig. 7.7.

If the dispersion-compensating fiber has a large negative dispersion, a length shorter than the link fiber would be needed to compensate for the accumulated dispersion. Thus, if the link fiber has a dispersion value of +17 ps/km·nm and

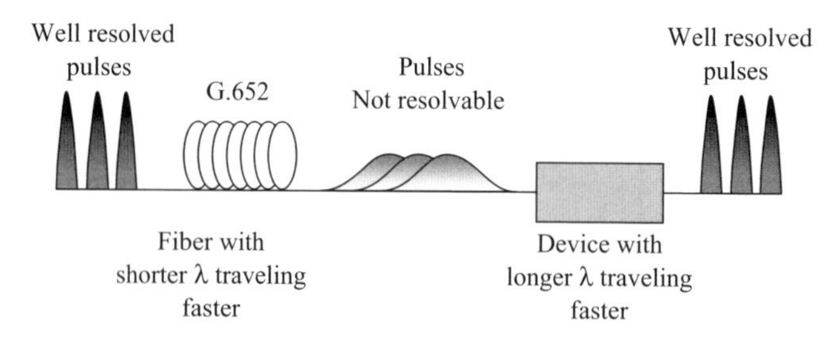

FIGURE 7.7 Concept of dispersion compensation. The link fiber is characterized by positive dispersion (i.e., longer wavelengths travel more slowly in the link fiber). The dispersion can be compensated by using a fiber in which longer wavelengths travel faster than shorter wavelengths.

the link is 100 km in length, the accumulated dispersion would be $(17)(100) = 1700$ ps/nm. If the dispersion-compensating fiber has a dispersion value of -170 ps/km·nm, to compensate for $+1700$ ps/km·nm, a length of only 10 km would be required. This led to the development of special fiber designs capable of having large negative dispersion in the 1550-nm wavelength window.

Figure 7.8 shows a schematic of two typical refractive index profiles of a dispersion-compensating fiber. By an appropriate choice of various parameters such as refractive indices and radii, it is possible using such designs, to achieve large negative dispersion. These dispersion-compensating fibers are usually placed in a spool at a site, along with optical amplifiers. Thus, although the signal propagates through the length of 10 km of dispersion-compensating fiber, this distance is not actually covered on the ground. In this context, fibers spans with alternating but equal-length segments of positive and negative dispersion fiber have been developed. In this case compensation of dispersion takes place as the optical pulses travel along the link itself. Such a technique is referred to as *dispersion management*, and such dispersion-managed fiber spans have been used to transmit terabits of information over undersea links.

With the demand for very high negative dispersion fibers, new fiber designs have evolved. A refractive index profile, which is characterized by a large dispersion coefficient ≈ -1800 ps/km·nm at 1550 nm, is shown in Fig. 7.9. A short length of such a DCF can be used in conjunction with the 1310-nm optimized fiber link so as to have a small total dispersion value at the end of the link.

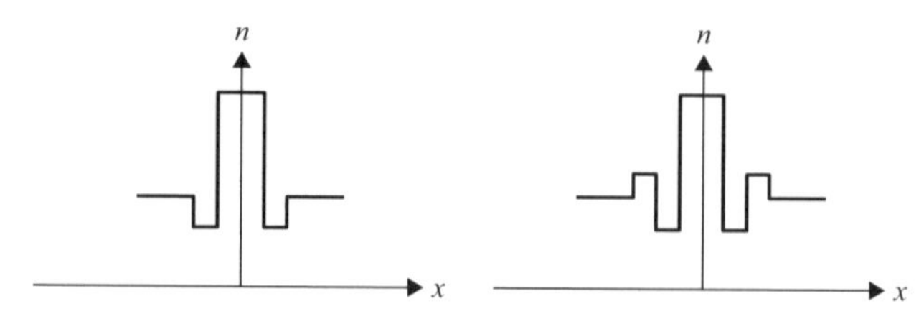

FIGURE 7.8 Typical refractive index profiles of dispersion-compensating fibers.

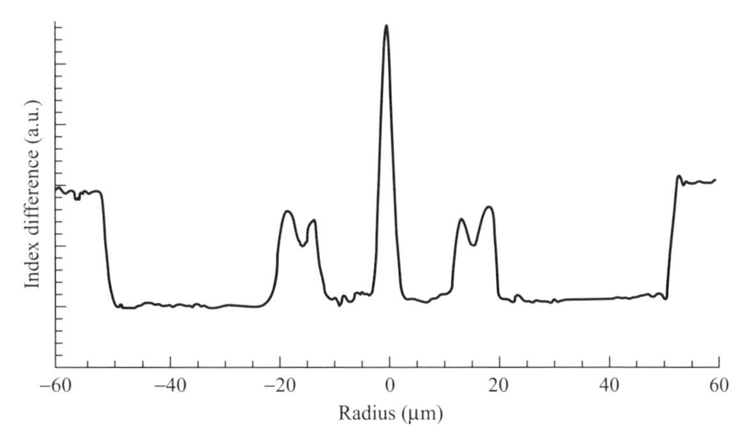

FIGURE 7.9 By proper modification of the transverse refractive index profile, it is possible to achieve very high negative dispersion values. The experimentally realized profile shown exhibits a dispersion coefficient of −1800 ps/km·nm at 1550 nm. (Adapted from Auguste et al., 2000.)

The dispersion of a given fiber depends on the wavelength. Thus, a link of 100 km of G.652 fiber operating from a source with a spectral width of 0.1 nm would have dispersion values of about 159, 165, and 171 ps at 1530, 1540, and 1550 nm, respectively. If a wavelength-division multiplexed system operating, say, with wavelengths of 1530, 1540, and 1550 nm is to operate efficiently, it is necessary to compensate for the accumulated dispersion at all three wavelengths. Hence, the dispersion-compensating fiber would also have to have the right wavelength dependence so that all the multiple wavelengths can be compensated simultaneously using a single dispersion-compensating fiber. Such dispersion-compensating fibers, called *dispersion slope–compensating fibers*, are now available for different types of fibers used commercially.

In recent years, DWDM (dense wavelength-division multiplexed) systems are often used in the wavelength region 1530 to 1565 nm, which represents the gain window of erbium-doped fiber amplifiers (EDFAs) (see Chapter 9). Due to simultaneous transmission at many closely spaced wavelengths and high optical gain from an EDFA, nonlinear effects like such as four-wave mixing impose serious limitations, especially if a dispersion-shifted fiber with a zero-dispersion wavelength at 1.55 μm is used (see Chapter 13). To overcome this difficulty, the use of small dispersion fiber has been suggested, where the dispersion is typically in the range 2 to 8 ps/km·nm (Table 7.1). Because of this, the phase-matching condition for efficient four-wave mixing is not satisfied, hence the effect of FWM becomes negligible. Thus, using DWDM in a small residual dispersion fiber together with EDFA and DFB laser diodes, one can get very high bit rates over a few hundred kilometers without using a repeater. However, if one wants repeaterless transmission over very large distances, the residual dispersion (2 to 8 ps/km·nm) in these fibers will go on accumulating and will limit the bit rate at each wavelength. To overcome this difficulty, one has to

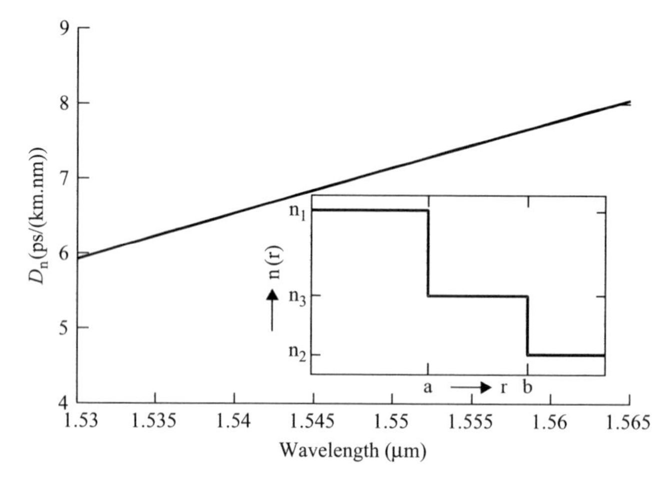

FIGURE 7.10 Variation of the total dispersion (D_N) of an SRDF as a function of wavelength. The refractive index profile is shown in the inset. (Adapted from Goyal et al., 2003.)

use a DCF that will compensate for the accumulated dispersion at all wavelengths simultaneously. The design of DCFs will therefore have to be compatible with the small residual dispersion fibers.

In the inset of Fig. 7.10, we have given typical refractive index variations of a small dispersion fiber called a *small residual dispersion fiber* (SRDF). The figure also shows the corresponding total dispersion (D_N) as a function of wavelength. In the inset of Fig. 7.11, we have given typical refractive index variation of the corresponding

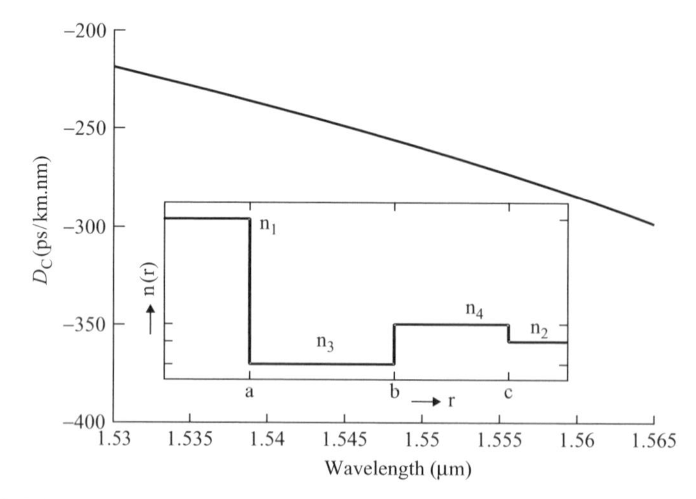

FIGURE 7.11 Variation of the total dispersion (D_C) of a DCF as a function of wavelength. The refractive index profile is shown in the inset. (Adapted from Goyal et al., 2003.)

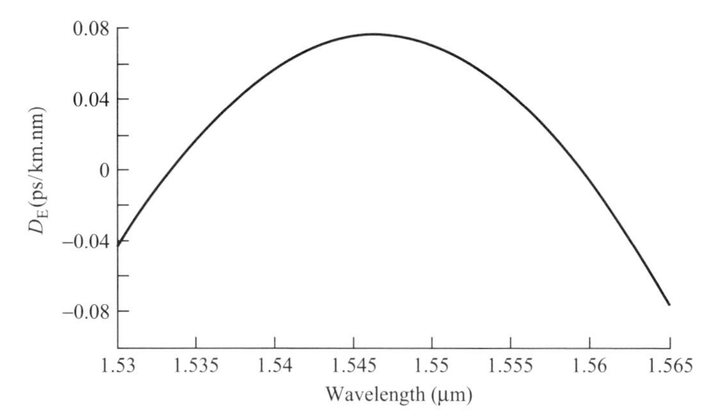

FIGURE 7.12 Variation of the effective dispersion (D_E) of the system as a function of wavelength. (Adapted from Goyal et al., 2003.)

DCF; the corresponding wavelength dependence of the total dispersion (D_C) has also been shown. The dispersion slopes are so adjusted that a small DCF length will compensate approximately for the accumulated dispersion in SRDF simultaneously at all wavelengths. In Fig. 7.12 we have plotted the net dispersion coefficient of a link consisting of length L_1 of SRDF and length L_2 of the corresponding dispersion-compensating fiber:

$$D_E = \frac{L_1 D_N + L_2 D_C}{L_1 + L_2}$$

for $L_1 = 36.74\, L_2$, where D_N and D_C represent the dispersions associated with SRDF and DCF, respectively. It may be noted that the maximum value of the effective dispersion is less than 0.08 ps/km·nm over the entire C-band.

7.4 POLARIZATION MODE DISPERSION

Normally, single-mode fibers used in communication are supposed to be circular in cross section. Thus, if we launch a linearly polarized beam in such a fiber, the polarization state of the light beam should not change as it propagates. In practice, it is found that the state of polarization is not maintained as light propagates in the fiber, and the output state of polarization is in general arbitrary. This change of state of polarization of the light happens due to many factors, such as slight ellipticity of the core, uneven stress distributions in the fiber when the fibers are manufactured, or bends and twists when the fiber is laid on the table or ground. To understand how this can affect light propagation, let us consider light propagation through an elliptic core fiber.

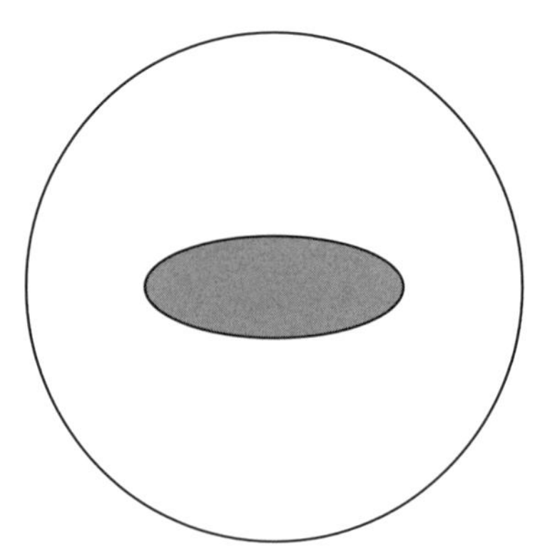

FIGURE 7.13 Elliptic core fiber.

Figure 7.13 shows an elliptic-core single-mode fiber in which the core, rather than being circular in cross section, has an elliptical geometry. Due to the geometry of this fiber, it is found that if we launch linearly polarized light that is polarized along the major or minor axis, the output from the fiber has the same polarization as the input and the fiber maintains this polarization. At the same time the light beam launched in a polarization parallel to the major axis of the ellipse travels with a velocity that differs from the light launched with its polarization parallel to the minor axis. Such fibers, referred to as *polarization-maintaining fibers*, are used in many applications, such as fiber optic gyroscopes (see Chapter 14).

Now if we launch a light pulse in an elliptic-core fiber with an arbitrary polarization state, the component of the input pulse with polarization parallel to the major axis will propagate at a different speed than that of the component parallel to the minor axis. This would result in a situation such as that shown in Fig. 7.14, wherein at the

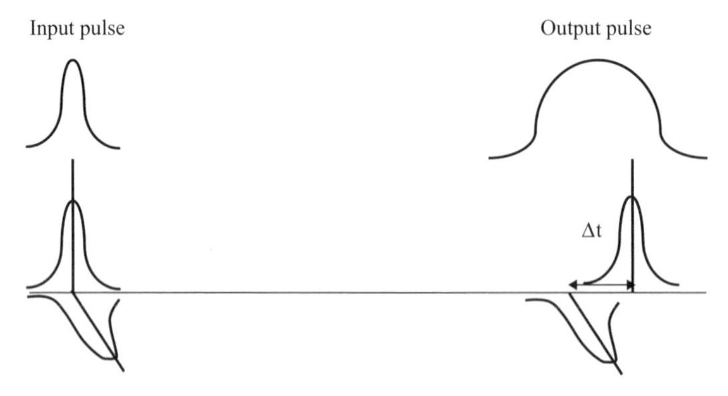

FIGURE 7.14 Components parallel to the major axis of the ellipse of an elliptic core fiber travel at a different velocity than do components parallel to the minor axis. This results in a separation of components at the output end and leads to polarization mode dispersion.

output the two polarization components arrive at slightly different times. Hence, the output pulse will be broader than the input pulse, and this dispersion phenomenon is referred to as *polarization mode dispersion*.

In circular-core fibers, due to various manufacturing defects or due to the particular way the fibers are laid in the link, similar effects can take place due to bends and uneven stresses in the fiber. Thus, the orientations of the two axes (the major and minor axes) are now distributed randomly along the length of the fiber. Hence, unlike an elliptic-core fiber, the differential velocity among the two polarization components is random along the length of the fiber. This effect still results in different times taken by two different polarization components and leads to polarization mode dispersion. In a practical fiber optic system, polarization mode dispersion changes randomly with time, due to variations in stress along the fiber because of temperature variations, vibrations of passing vehicles, and other factors. Due to this temporal dependence, the fiber optic link could be working perfectly at certain times but fail at certain other times, due to increased polarization mode dispersion. Due to the complete random nature of the polarization mode dispersion problem (similar to random walk), this dispersion is specified in units of $\text{ps}/\sqrt{\text{km}}$. Thus, if the length of the fiber is increased fourfold, the polarization mode dispersion would only increase by a factor of 2.

For a communication system working at 2.5 Gb/s (with pulses of duration 400 ps), the maximum allowed value of PMD is about 40 ps, while for a 40-Gb/s (with pulses of width 25 ps) system, the maximum PMD allowed is only 2.5 ps. Thus, PMD effect becomes more and more important with increasing bit rates. Techniques are now available to compensate for PMD effects which are important for high-bit-rate systems. Also, with the improvements in technology of fiber manufacture, PMD values have been reduced significantly in commercially available fibers. Commercially available single-mode fibers typically have a PMD of less than $0.2\,\text{ps}/\sqrt{\text{km}}$. In an actual system, contributions to PMD also come from other components used in the link, such as optical amplifiers and dispersion compensating elements.

CHAPTER 8

Fiber Optic Communication Systems

8.1 INTRODUCTION

As we discussed in Chapter 3, in optical fiber communication systems, the information is coded in the form of optical pulses (a 1 or a 0) which propagate along optical fiber links. In a point-to-point long-distance optical fiber communication link, these optical pulses would have to propagate over very long distances, such as hundreds to thousands of kilometers. We know that as optical pulses carrying information propagate through an optical fiber, they get attenuated and lose power and get broadened in time due to dispersion. For retrieving information, the optical pulses need to be detected and converted to electrical signals for further processing. Optical detectors need to receive a minimum optical power to be able to decipher the sent bit, whether it is a 1 or a 0. Also, if the pulse dispersion is large, the adjacent pulses may start to overlap, resulting in nonresolvable pulses and leading to errors in detection.

To propagate over long distances, an actual system uses *regenerators*, which are placed periodically along the link and compensate for the accumulated loss and dispersion (Fig. 8.1). In the case of electronic regenerators, the incoming optical pulses are first converted into electrical pulses, which are then processed in the electronic domain to retime and reshape the pulses. This process also removes any noise that may have accumulated in the pulses. The resulting electrical pulses are then amplified and used to drive a laser diode, resulting in a fresh optical pulse stream. In this way the pulse stream leaving the regenerator is almost as good as it was when it started from the transmission end. When loss and dispersion accumulate due to further propagation, another regenerator compensates, and in this way the information-carrying pulse stream is able to propagate over very long distances without much accumulation of errors.

If an optical communication system employs wavelength-division multiplexing with multiple signal wavelengths carrying information, at each regenerator site we would first need to demultiplex (separate) the various channels, use as many regenerators as the number of wavelengths, and after regenerating the channels would need to be multiplexed (combined) into a single output for further transmission. Such electronic regenerators would be very expensive solutions for WDM systems.

Fiber Optic Essentials, By K. Thyagarajan and Ajoy Ghatak

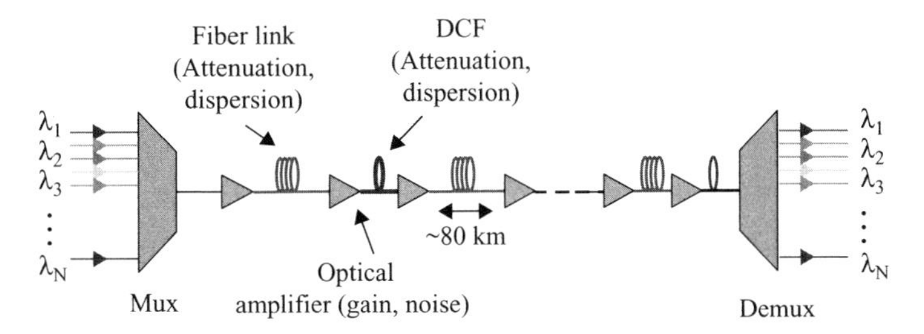

FIGURE 8.1 Long-haul wavelength-division-multiplexed fiber optic communication system with amplifiers and dispersion compensators spaced out along the link. Mux: multiplexer; demux: demultiplexer.

The accumulated loss can indeed be compensated for by optical amplifiers, which amplify the optical signals in the optical domain itself without conversion to electrical pulses; these are discussed in detail in Chapters 9 and 10. Indeed, a single optical amplifier can amplify signals at multiple wavelengths and thus would be much more economical than electronic regenerators. However, the optical amplifiers would not be able to retime and reshape the signal pulses and would add noise to the pulse stream. This added noise would finally determine the number of amplifier sections that the signals can propagate along. Since optical amplifiers would not compensate for dispersion, as discussed in Chapter 7 the accumulated dispersion would have to be compensated optically using dispersion compensators. The number of sections of optical amplifiers and dispersion compensators that can be employed in a link would be determined by the accumulated noise and any other distortion, including that due to nonlinear effects. If the pulses need to travel further, they would need to be regenerated electronically.

In this chapter we look at some of the design issues of optical fiber communication systems. First we look very briefly at optical sources and detectors and then discuss the issues of budgeting for power and dispersion. For a detailed account of various practical issues in optical networking, readers may look up Ramaswami and Sivarajan (1998).

8.2 OPTICAL SOURCES

The objective of an optical transmitter is to generate optical pulses corresponding to the information and then couple this light into the optical fiber. The light sources used in optical fiber communication systems are either light-emitting diodes (LEDs) or laser diodes (LDs). Both sources are based on semiconductor materials and emit light when current is made to pass through them. Using different semiconductors it is possible to achieve emission at different wavelengths. Lasers used at 1310 and 1550 nm are usually based on a compound semiconductor: gallium indium arsenide phosphide (GaInAsP).

Laser Diodes

The *semiconductor laser* (also referred to as a *junction laser* or *diode laser*) is today one of the most important types of lasers for application in fiber optic communication. These lasers use semiconductors as the lasing medium and are characterized by specific advantages, such as the capability of direct modulation in the gigahertz region, small size and low cost, the capability of monolithic integration with electronic circuitry, direct pumping with conventional electronic circuitry, and compatibility with optical fibers.

The basic mechanism responsible for light emission from a semiconductor is the recombination of electrons and holes at a *p–n* junction (a junction between a *p*-type semiconductor and an *n*-type semiconductor) when a current is passed through a diode. Just as in other laser systems, there can be three interaction processes: (1) an electron in the valence band can absorb the incident radiation and be excited to the conduction band, leading to generation of an electron–hole pair; (2) an electron can make a spontaneous transition in which it combines with a hole (i.e., it makes the transition from the conduction to the valence band and in the process, emits radiation); and (3) a stimulated emission may occur in which the incident radiation stimulates an electron in the conduction band to make a transition to the valence band and in the process emit radiation. If, by some mechanism, a large density of electrons is created at the bottom of the conduction band and simultaneously, in the same region, a large density of holes is produced at the top of the valence band (Fig. 8.2) by the process of stimulated emission, such a semiconductor can amplify optical radiation at a frequency that corresponds to energy slightly greater than the bandgap energy. To convert the amplifying medium into a laser, one must provide optical feedback, which is usually done by cleaving or polishing the ends of the *p–n* junction diode at right angles to the junction (Fig. 8.3). Such lasers are also referred to as *Fabry–Perot laser diodes*.

Thus, when a current is passed through a *p–n* junction under forward bias, the injected electrons and the holes will increase the density of electrons in the conduction

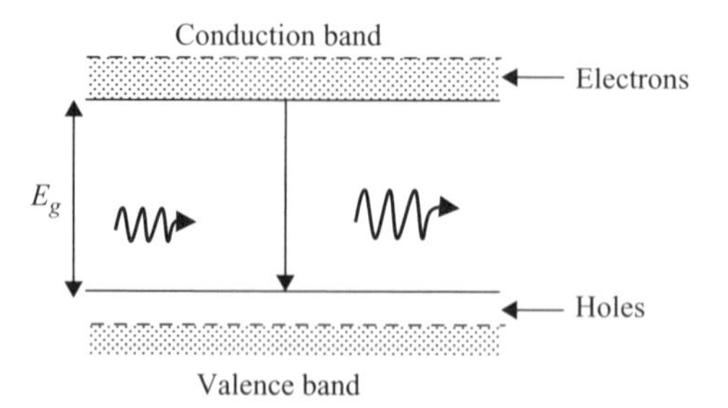

FIGURE 8.2 If a large density of electrons is created at the bottom of the conduction band and simultaneously in the same region a large density of holes is produced at the top of the valence band, such a semiconductor can amplify optical radiation at a frequency that corresponds to energy slightly greater than the bandgap energy.

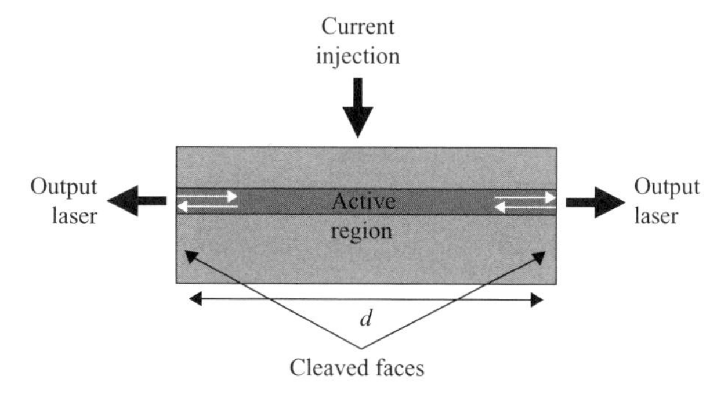

FIGURE 8.3 Fabry–Perot laser diode. The cleaved end faces form the mirrors for the laser cavity.

band and the holes in the valence band at a region close to the junction and at some value of current, the stimulated emission rate will exceed the absorption rate and amplification will begin. As the current is increased further, at some threshold value of current, the amplification will overcome the losses in the cavity and the laser will begin to emit coherent radiation.

Early semiconductor lasers were based on *p–n* junctions formed on the same material by proper doping, referred to as *homojunction lasers*. Due to the absence of potential barriers for the confinement of carriers or abrupt refractive index discontinuities for the confinement of optical radiation, these laser structures required large thereshold current densities (about 50,000 A/cm^2). The absence of carrier confinement resulted in a diffusion of the carriers near the *p–n* junction plane, due to which a significant optical gain was available only over a very small region around the junction. The absence of any strong optical confinement resulted in the optical energy penetrating beyond the gain region, where it was absorbed. Thus, larger current densities were required for laser operation.

A significant reduction in threshold current densities was achieved by forming *heterojunctions*, junctions formed between two dissimilar semiconductors. Present-day lasers are based on a double heterojunction in which a thin active layer of a semiconductor with a narrow bandgap is sandwiched between two larger bandgap semiconductors as shown in Fig. 8.4. In this configuration, the regions in which the electrons and holes recombine is bound on either side by potential barriers, and thus they are confined to the thin active region. Fortuitously, the refractive index of the semiconductor decreases with an increase in the bandgap. Thus, the refractive index of the central active layer is higher than that of the two surrounding regions (typically, about 5 to 10% higher). By the mechanism of waveguidance, which occurs due to total internal reflections taking place at the boundaries, such a refractive index profile can confine the emitted optical radiation to the active region. In addition, since the layers surrounding the active central region have larger bandgaps, the optical field that penetrates the surrounding region is also not absorbed. Use of double heterojunctions results in a reduction in the threshold current densities to about 2000 to 4000 A/cm^2. Taking a typical cavity length of 300 μm and a width of 100 μm, the threshold current required

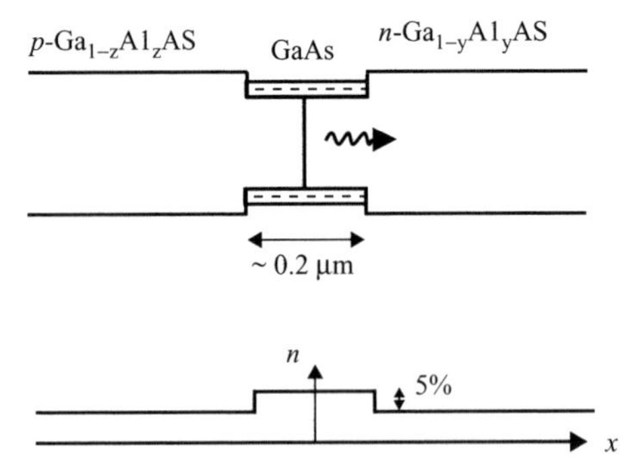

FIGURE 8.4 Energy diagram of a double heterostructure laser. The lower figure gives the refractive index distribution.

for a threshold current density of 4000 A/cm^2 will be about 1.2 A. Figure 8.5 shows a buried heterostructure laser in which strong lateral confinement is provided through a large refractive index step using different semiconductor materials. The active region has typical dimensions of 0.1 μm by 1 μm, and the refractive index step is about 0.2 to 0.3. Because of the strong light confinement provided by the index step, such lasers operate in a single transverse mode and are stable against current variations.

Most semiconductor lasers operate either in the wavelength region 0.8 to 0.9 or 1 to 1.7 μm. Since the wavelength of emission is determined by the bandgap, different semiconductor materials are used for the two wavelength regions. Lasers operating in the spectral region 0.8 to 0.9 μm are based on gallium arsenide (GaAs). By replacing a fraction of gallium atoms by aluminum, the bandgap can be increased. Thus, one can form heterojunctions by proper combinations of GaAlAs and GaAs which can provide both carrier confinement and optical waveguidance. For example, the bandgap of GaAs is 1.424 eV and that of $Ga_{0.7}Al_{0.3}As$ is ≈1.798 eV; the corresponding refractive index difference is about 0.19. Thus, surrounding the GaAs layer on either side with $Ga_{0.7}Al_{0.3}As$, one can achieve confinement of both carriers and light waves.

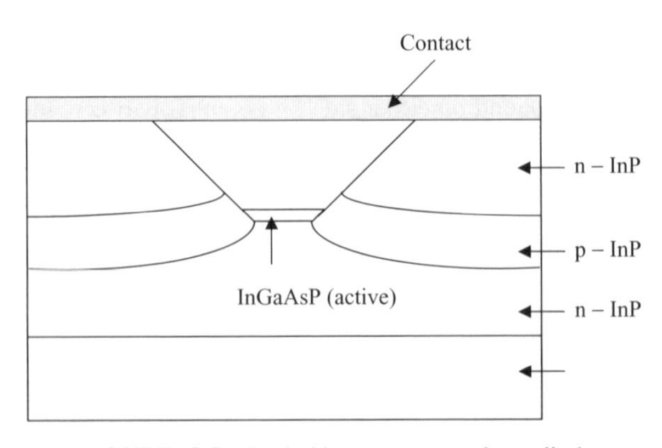

FIGURE 8.5 Buried heterostructure laser diode.

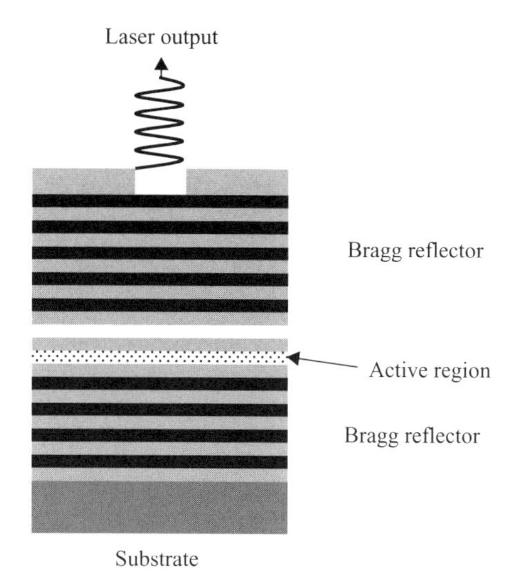

FIGURE 8.6 Vertical cavity surface-emitting laser.

For lasers operating in the wavelength band 1.0 to 1.7 μm, the semiconductor material is InP, with gallium and arsenic used to replace fractions of indium and phosphorus, respectively. This wavelength region is extremely important in connection with fiber optic communication, since silica-based optical fibers exhibit both low loss and very high bandwidth around 1.55 μm.

A new class of lasers, *vertical cavity surface-emitting lasers* (VCSELs, pronounced "vicsels"), emit light vertically from the laser surface. The laser cavity is vertical, whereas it is horizontal in conventional laser diodes. Figure 8.6 shows a typical VCSEL structure, with the active region sandwiched between two multilayer mirrors with as many as 120 mirror layers. The multilayer mirrors act like a Bragg grating and reflect only a narrow range of wavelengths back into the cavity, causing emission at a single wavelength. Compared to conventional edge-emitting laser diodes, it is possible to realize an array of VCSELS on a surface, and they are also very small in size and have threshold currents below 1 mA.

Figure 8.7 shows the output spectrum from a laser diode below threshold when no lasing is taking place and the output is only spontaneous emission, and above threshold when the device starts to lase. Figure 8.8 shows a typical light output versus current characteristic of a laser diode. As can be seen, the output optical power starts to increase very rapidly around a threshold current, which essentially represents the beginning of laser oscillation. In digital modulation of laser diodes, they are biased at slightly above threshold, and on this bias is superposed current pulses corresponding to the digital data. Thus, the electrical signal can be encoded directly into an optical signal. For analog modulation the laser is usually biased above threshold, and the analog signal is fed in the form of current variations.

An important characteristic of a laser diode is the spectral width of emission. Figure 8.9 shows typical output spectra of a single longitudinal mode and a multilongitudinal

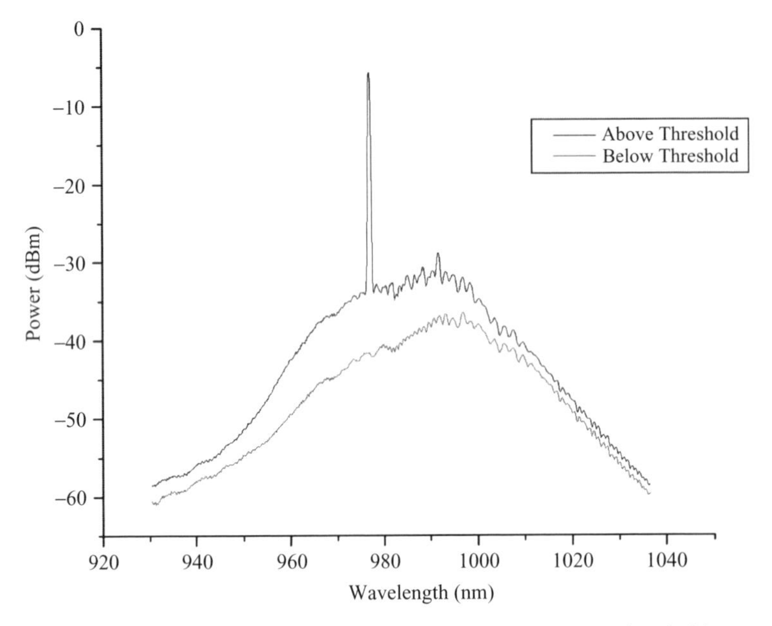

FIGURE 8.7 Output from a laser diode below and above threshold.

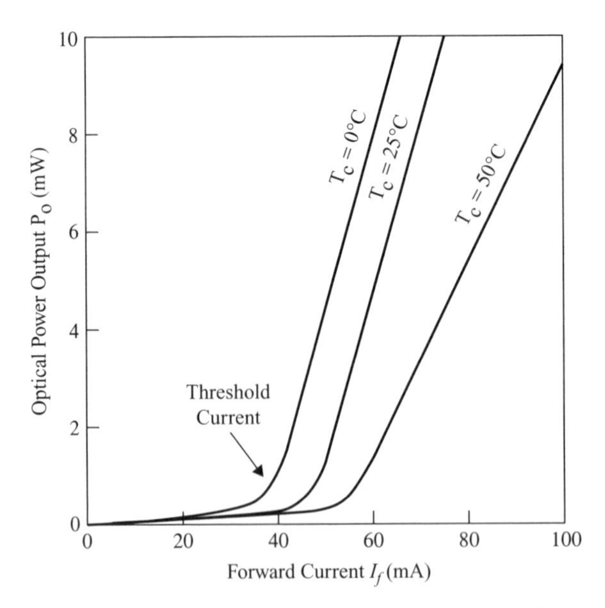

FIGURE 8.8 Typical output power versus current in a laser diode. (Adapted from http://www.fiber-optics.info/articles/laser-diode.htm.)

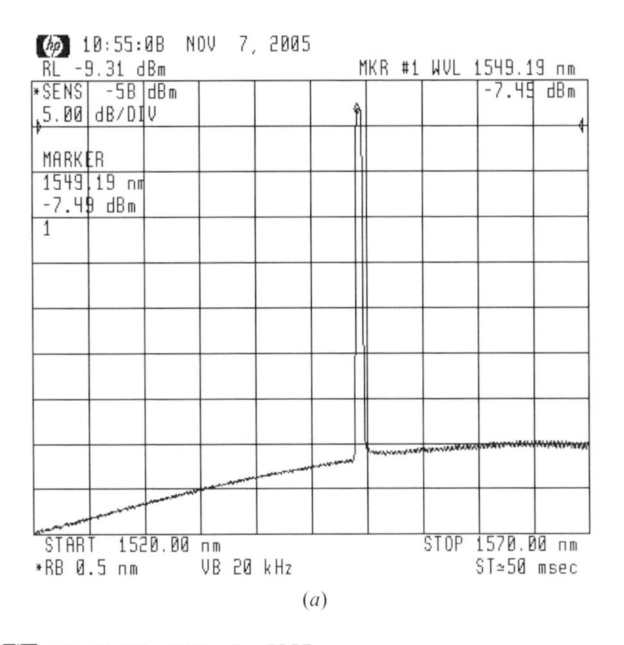

(*a*)

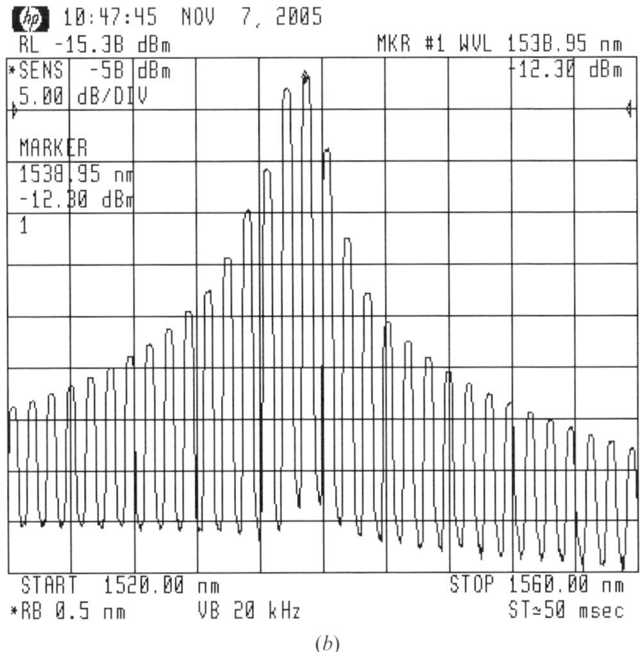

(*b*)

FIGURE 8.9 Typical output spectrum of (*a*) a single longitudinal mode laser and (*b*) a multiple longitudinal mode laser.

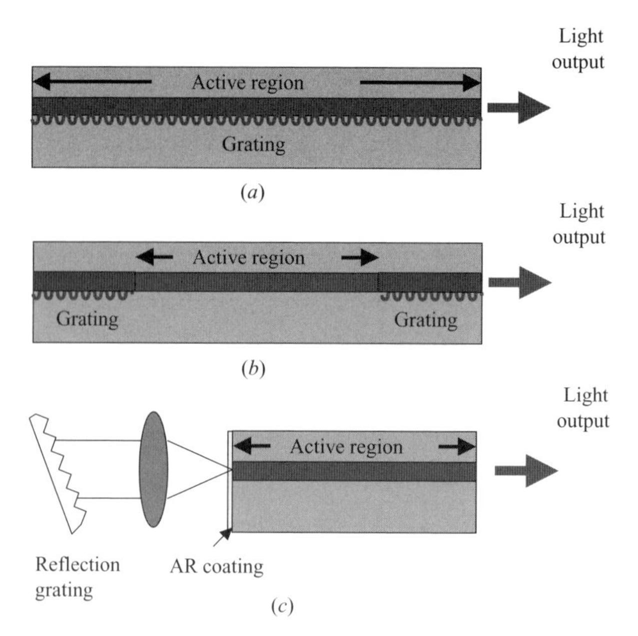

FIGURE 8.10 (*a*) Distributed feedback laser; (*b*) distributed Bragg reflector laser; (*c*) external cavity laser.

mode laser. In some applications (e.g., in fiber optic communications) one would like to have single longitudinal mode oscillation of the laser so that its spectral width is very small ($\ll$0.1 nm). One can achieve this by using distributed Bragg reflectors or distributed feedback. The distributed Bragg reflection is achieved using periodic variation of the thickness of the layer surrounding the active region of the laser (Fig. 8.10*a*). This essentially acts as a Bragg reflector which is highly wavelength selective (see Chapter 11). In a distributed Bragg reflector laser (DBR), the Bragg reflectors are present at the ends of the cavity (Fig. 8.10*b*), whereas in a distributed feedback (DFB) laser, the Bragg reflectors are distributed throughout the length of the cavity. Since reflection by Bragg structure is highly wavelength selective, this leads to single longitudinal mode oscillation of the laser. By providing an antireflection coating on one of the surfaces and providing for reflection from an external mirror, it is possible to tune the laser wavelength; such lasers are termed *external cavity lasers* (Fig. 8.10*c*).

The emission from laser diodes is much more directional than an LED, and hence much higher coupling efficiencies into single-mode fibers are possible, leading to higher optical power levels for transmission over longer distances. LDs also have very narrow spectral widths, leading to much reduced pulse dispersion and hence much higher information-carrying capacities. Figure 8.11 shows a comparison of the emission from a laser diode and an LED. Hence, LDs are the choice of sources for long-distance, high-speed communication using single-mode fibers.

When LDs are used in WDM systems, their wavelengths need to be set precisely at specified wavelengths following International Telecommunications Union (ITU)

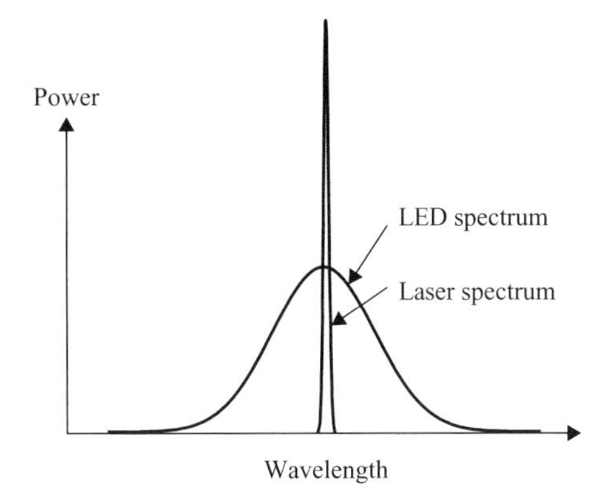

FIGURE 8.11 Comparison of the output spectrum from a laser diode and from an LED.

specifications. Table 8.1 gives the standard frequencies and corresponding wavelengths. Note that the channels are equally spaced in frequency, but their wavelength spacing is not the same. The frequency spacing given is 1 THz; dense wavelength-division-multiplexing (DWDM) schemes operate with a frequency spacing of 100 or 50 GHz. Since WDM systems operate with very closely lying wavelengths (separated by only 0.8 nm, which is approximately 100 GHz at a wavelength of about 1550 nm), there should also be no drift in the wavelengths of individual lasers. This is achieved using external cavity lasers wherein feedback from one side is provided by an external reflector that is highly wavelength selective, like a fiber Bragg grating (Fig. 8.12).

TABLE 8.1 ITU Grid Frequencies and Wavelength

Frequency (THz)	Wavelength (nm)
186	1611.78
187	1603.16
188	1594.64
189	1586.20
190	1577.85
191	1589.59
192	1561.41
193	1553.32
194	1545.32
195	1537.39
196	1529.55
197	1521.78
198	1514.10
199	1506.49
200	1498.96

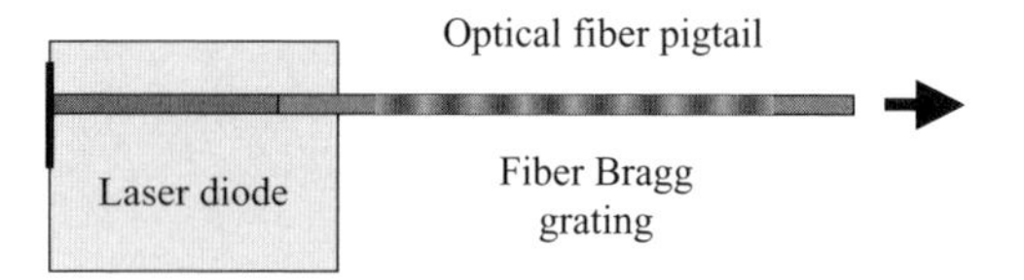

FIGURE 8.12 External cavity laser diode with a fiber Bragg reflector as a wavelength-selective reflector.

Since the reflection occurs at a very precise wavelength, the laser oscillates at the wavelength fixed by the external reflector.

Some designs of LDs also provide for tuning of their wavelengths. Since the lasing wavelength depends on the temperature of the laser diode, one of the techniques used is temperature tuning. Lasers with tuning over a wavelength range of about 35 nm (1528 to 1563 nm) with output powers of 10 mW are available commercially. The spectral widths of these lasers can also be very small (about 5 MHz). Other techniques use external cavity with a Fabry–Perot etalon for tuning the laser wavelength over the entire C-band (191.7 to 196 THz). Figure 8.13 shows a typical laser output spectrum of a tunable laser that has an output power of +13 dBm (=20 mW). Figure 8.14 shows a laser chip consisting of a laser diode, semiconductor optical amplifier, and an external modulator called an electroabsorption modulator. Figure 8.15 shows composite optical spectra of 20 tuned wavelength positions spaced by 50 GHz, with peak power of 1 mW (0 dBm), showing the tunability and high power achievable.

The choice of laser diodes in a communication system depends on the operating wavelength, power output, spectral width, and modulation speed, the last being how fast a laser can be modulated to generate a pulse stream. The output power of an LD is determined by the current passing through it. Thus, if the current is modulated, it is possible to achieve direct modulation of the light. It is thus possible to generate the bit sequence easily using LDs by direct modulation. Of course, the modulation

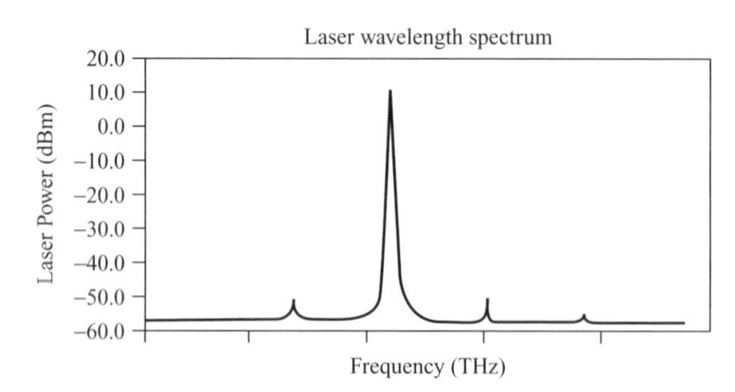

FIGURE 8.13 Output spectrum of a tunable diode laser. (Adapted from *Intel C-band Tunable laser*, Performance and Design, White Paper, May 2003.)

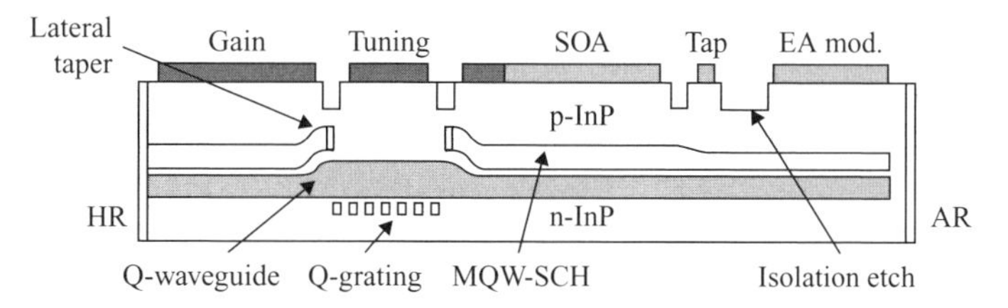

FIGURE 8.14 Tunable laser diode with a semiconductor optical amplifier external electro-absorption modulator integrated with the laser. (After Johnson et al., 2001. Copyright © 2001 IEEE.)

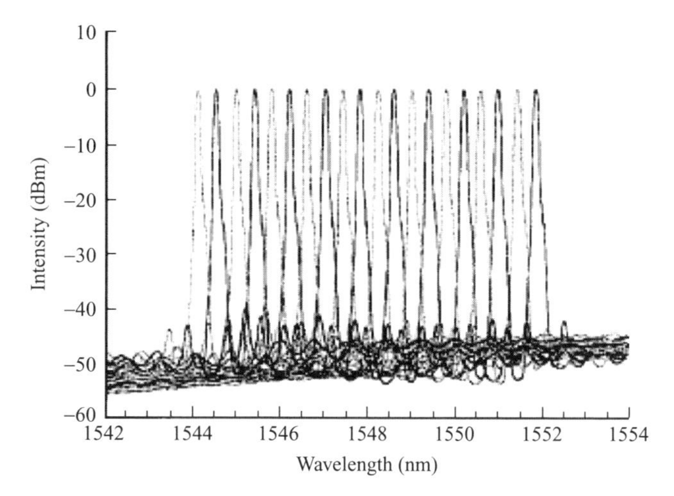

FIGURE 8.15 Composite optical spectra of 20 tuned wavelength positions spaced by 50 GHz and with peak power of 1 mW (0 dBm). (After Johnson et al., 2001. Copyright © 2001 IEEE.)

bandwidth of the LDs should be sufficiently large to achieve the speed of modulation desired.

If the end surfaces are antireflection coated or angle polished so that there are no reflections from the end faces back into the cavity, then when current is passed through the device, it will function as an amplifier rather than as a laser. In fact, this is the principle of semiconductor optical amplifiers. In such cases the reflectivities from the end faces is brought to less than 0.01%, so that the device does not oscillate but acts like an amplifier.

Light-Emitting Diodes

Compared to laser diodes, in the case of LEDs the emission comes primarily from spontaneous emissions (see Section 9.2 for definition of spontaneous emission), and hence the light output appears over a large angle and the emission is over a broad band of wavelengths, typically about 30 to 50 nm wide (Fig. 8.16). Such large spectral widths lead to large pulse dispersion when used in a communication system.

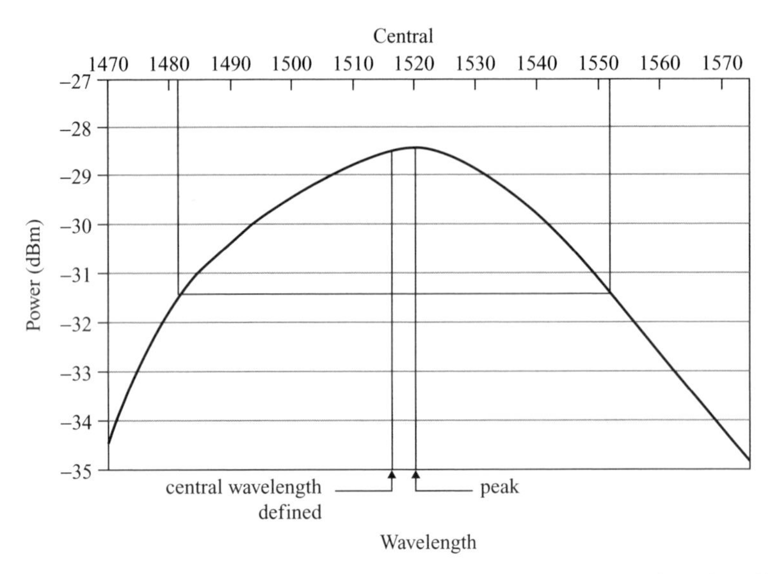

FIGURE 8.16 Output spectrum from a light-emitting diode. (Adapted from http://www.exfo.com/en/support/WaveReview/2005-February/WRarticle2.asp.)

Due to the emission pattern not being directional, the amount of light that can be coupled into a single-mode fiber is usually very small. However, efficient coupling into multimode fibers is possible using LEDs. The coupling efficiency of light from an LED to a multimode fiber depends on the fiber diameter and the numerical aperture of the fiber. For example, with an LED emitting 1 mW of power, one can approximately couple about 40 μW into a typical multimode fiber, whereas only about 10 μW can be coupled into single-mode fibers. Modulation speeds of LEDs are normally restricted to about 1 GHz. Due to the low bandwidth capabilities and the high beam divergence of the output beam, LEDs are used with multimode fibers for short-distance applications. Typical commercially available LEDs for operation at 1310 nm give power outputs of −20 dBm in 50-μm-core-diameter fibers (−17 dBm power in 62.5-μm-core-diameter fibers) with a spectral bandwidth of 100 to 170 nm.

Most long-distance communication systems use laser diodes as sources in transmitters. Since much higher power levels can be coupled into the fibers, most long-distance systems are based on single-mode fibers into which coupling of light from an LED is very inefficient, and laser diodes have a much smaller spectral bandwidth, resulting in much less pulse dispersion.

8.3 PHOTODETECTORS

The primary purpose of a photodetector is conversion of an input optical signal into an electrical signal for further processing. Like laser diodes, photodetectors used in fiber optic communication are also based on semiconductors. Detectors for optical

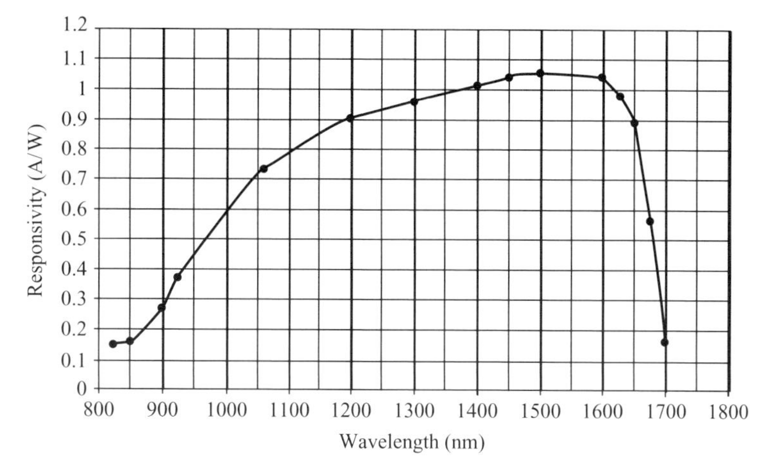

FIGURE 8.17 Typical wavelength dependence of the responsivity of an InGaAs photodiode.

communications in the 1310- and 1550-nm windows are based on the compound semiconductor Indium Gallium Arsenide (InGaAs); for shorter wavelengths, around 850 nm, the detectors are based on silicon or gallium arsenide.

When light falls on a detector current is generated, its magnitude depending on the power falling on the detector. The detector converts the optical pulse stream into an electrical pulse stream, which can be processed using electronic circuits. Typically, photodetectors produce 0.5 mA of current for an input optical power of 1 mW. The ratio of the current produced to the input optical power is referred to as *responsivity*. The photodetector described above would have a responsivity of 0.5 mA/mW. Figure 8.17 shows the typical responsivity of an InGaAs photodetector. The sensitivity peaks around 1550 nm, and such a photodetector is ideal for fiber optic communication systems operating in the 1550-nm window.

There are primarily two types of semiconductor detectors: *p*-type–intrinsic–*n*-type (PINs) and avalanche photo detectors (APDs). In the former there is no amplification within the detector, whereas in the latter case there is an avalanche multiplication of electrons within the detector itself, brought about by high electric fields within the detector.

Another important aspect of photodetectors is the speed of response. This is determined primarily by three factors: the transit time of the photo-generated carriers through the detector, the electrical frequency response as determined by the resistance and capacitance in the circuit (including that of the photodetector itself), and slow diffusion of the current carriers within the detector. To achieve high bandwidth, photodetectors should have a small capacitance, which in turn implies that they must have a small area. Typical capacitance values of photodetectors are about 5 pF. If the resistance in the circuit is 100 Ω, the rise time of the detector is about 0.9 ns.

In APDs an internal current gain by multiplication of the current carriers is achieved by providing a large reverse bias. The large electric field within the detection area

TABLE 8.2 Typical Performance Characteristics of Two Commonly Used Photodetectors

Parameter	Silicon PIN	Silicon APD	InGaAs PIN	InGaAs APD
Wavelength range (nm)	400–1100	400–1100	900–1700	900–1700
Responsivity (A/W)	0.6	77–130	0.63–0.8 at 1300 nm 0.75–0.97 at 1550 nm	
Gain	1	150–250	1	10–30
Dark current (nA)	1–10	0.1–1	1–20	1–5
Rise time (ns)	0.5–1	0.1–2	0.06–0.5	0.1–0.5

results in rapid acceleration of the charges, which then collide and create more charge carriers. These in turn accelerate and create further charge carriers. This results in an avalanche effect and leads to much higher currents in the output for the same input optical power. Typical bias voltages required for silicon APDs are about 250 V; APDs operating at higher wavelengths require about 20 to 30 V for their operation.

Table 8.2 gives typical performance characteristics of photodetectors used in fiber optic communication systems. In the photodetection process the light signal is converted into an electrical signal. This takes place by generating in the semiconducting material electrons and holes which then flow in the circuit to generate current. This generation is random, which introduces noise in the electrical signal being detected. If the noise becomes comparable to the signal-generated current, this may lead to errors in detection. There are two primary noise-generating mechanisms in the detection process: shot noise and thermal noise.

Shot Noise

Shot noise arises due to the fact that an electric current is made up of a stream of discrete charges, electrons, which are generated randomly when light falls on a photodetector. Thus, even if the light intensity falling on the photodetector is constant, the current would fluctuate, leading to shot noise. Since shot noise current arises due to random fluctuations, the average shot noise current would be zero. The electrical shot noise power will be related to the shot noise, and for this we define a mean-square shot noise current given by:

$$\bar{i}_{\mathrm{NS}}^2 = 2eI\,\Delta f \tag{8.1}$$

where e represents the charge of the electron ($= 1.6 \times 10^{-19}$ C), I is the average current produced due to the incident light beam, and Δf is the electrical bandwidth over which the noise is being considered. Since the photocurrent I depends on the incident optical power, the shot noise increases with increase in incident optical power. At the same time, the ratio of the photocurrent I_d to the shot noise current ($\sqrt{\bar{i}_{\mathrm{NS}}^2}$) increases with increase in input optical power, leading to greater discrimination between signal and noise at higher input optical powers.

Even in the absence of input optical power, a photodetector generates current due to thermal agitation of the carriers. This leads to *dark current noise* with an average noise current of zero. Since this is related to thermal agitation of the carriers, the noise term increases with increased temperature. Typical dark current noise for variously detectors are:

$$\begin{aligned} I_d &\sim 1 \text{ to } 10\,\text{nA} && \text{for silicon} \\ &\sim 50 \text{ to } 500\,\text{nA} && \text{for germanium} \\ &\sim 1 \text{ to } 20\,\text{nA} && \text{for InGaAs} \end{aligned}$$

In the presence of dark current, the total shot noise generated is given by

$$\bar{i}_{\text{NS}}^2 = 2e(I + I_d)\Delta f \tag{8.2}$$

Example 8.1 We consider an InGaAs detector with a responsivity of 0.5 A/W and a dark current of 5 nA. If the input optical power is 10 μW, the photocurrent generated will be 5 μA. If the detection bandwidth is 1 GHz, the root-mean-square shot noise current will be 40 nA, which is large compared to that of the dark current.

Thermal Noise

Thermal noise is generated as a result of the random motion of electrons brought about by the finite temperature of the system. Higher temperature leads to higher thermal energy of the electrons and hence to higher thermal noise. The mean-square thermal noise current in a load resistor R_L is given by

$$\bar{i}_{\text{NT}}^2 = \frac{4k_{\text{B}}T\,\Delta f}{R_L} \tag{8.3}$$

where k_B ($= 1.38 \times 10^{-23}$ J/K) is the Boltzmann constant, T is the absolute temperature, and Δf is the electrical bandwidth over which the noise is being considered. Thermal noise is independent of the input signal power and can be reduced by cooling the detector. The choice of R_L determines the bandwidth of the receiver. For larger bandwidths it is necessary to choose smaller values of R_L, resulting in larger noise.

Example 8.2 If we consider the detector used in Example 8.1, at a temperature of 300 K with a load resistor of 100 kΩ, the root-mean-square thermal noise current would be 13 nA. If the input signal power decreases, the shot noise would be reduced while the thermal noise remains the same. Thus, at high input signal powers, the shot noise term dominates, whereas at low input powers the thermal noise dominates.

Signal-to-Noise Ratio

The *signal-to-noise ratio* (SNR) is the ratio of the electrical signal power to the electrical noise power. As the SNR decreases, errors in detection increase. Thus, to keep the error rate below a specified value (typically, 10^{-12}, i.e., less than one error in 10^{12} pulses received), it is important to have an SNR greater than a certain value or, equivalently, an incident optical power above a certain threshold. The value of incident optical power required also depends on the bit rate and the type of detector used.

If the input optical power is P, the current generated is RP, where R is the responsivity of the detector. The electrical signal power would be proportional to the square of the current and hence to R^2P^2. The total noise power would be given by the sum of shot noise, dark current noise, and thermal noise; these noise powers will be proportional to the respective mean-square noise currents given by Eqs. (8.2) and (8.3), Hence, the SNR will be given by

$$\text{SNR} = \frac{R^2P^2}{2e(I + I_d)\Delta f + 4k_\text{B}T\,\Delta f/R_L} \tag{8.4}$$

In this equation, one of the noise terms usually dominates, depending on the input optical power. Under shot noise–limited operation (high input powers), one can neglect the thermal noise contribution, and under thermal noise–limited operation (low input powers), the shot noise term can be neglected.

We assume that the minimum detectable optical power occurs when the signal power and the noise power are equal. This optical signal power, referred to as the *noise equivalent power* (NEP), is usually stated in units of $\text{W}/\sqrt{\text{Hz}}$. Assuming that the minimum power will be so low that $I_d \gg I$, we obtain an expression for NEP by putting SNR = 1 in Eq. (8.4):

$$\text{NEP} = \frac{1}{R}\left(2eI_d + \frac{4k_\text{B}T}{R_L}\right)^{1/2} \qquad W/\sqrt{\text{Hz}} \tag{8.5}$$

Example 8.3 We consider a silicon photodetector with a responsivity of 0.65 A/W, $R_L = 1000\ \Omega$, $\Delta f = 100$ MHz, and $P = 0.5\ \mu$W. In such a case, neglecting the dark current noise, we see that the noise is limited by thermal noise, and we obtain a SNR of 18 dB. If the power is now increased to 0.5 mW, the detection is limited by shot noise and we obtain a SNR of 69 dB.

The speed of response of a photodetector is also a very important parameter and determines whether the detector would be able to follow rapid variations of power in the pulses received. Thus, for the photodetector to respond to signals at 2.5 Gb/s, the detector should be able to respond to changes within a time scale of about 100 ps.

Typical sensitivities at 1550 nm of PIN detectors at bit rates of 2.5 and 10 Gbit/s are −23 and −18 dBm, while those of APDs for these bit rates are −34 and

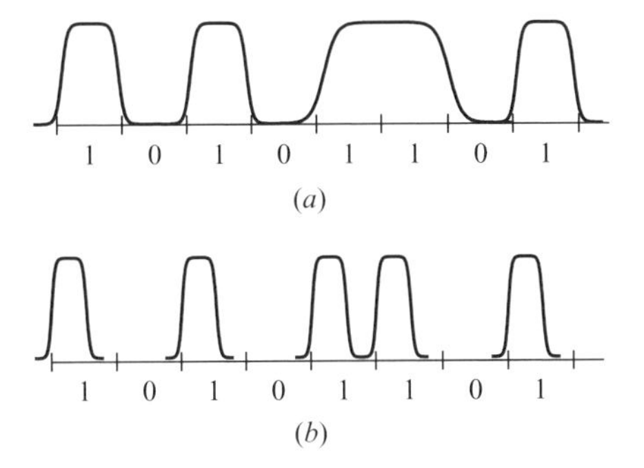

FIGURE 8.18 (*a*) NRZ and (*b*) RZ bit patterns corresponding to the bit sequence 10101101.

−24 dBm, respectively, showing the higher sensitivity of APDs. Further improvement in sensitivity is possible by the use of optical amplifiers.

8.4 NON-RETURN-TO-ZERO AND RETURN-TO-ZERO FORMATS

As mentioned earlier, the bit 1 is represented by a light pulse and the bit 0 by the absence of a pulse. The two commonly used formats for representing the digital pulse train are non-return-to-zero (NRZ) and return-to-zero (RZ) formats. For NRZ, the duration of each pulse is equal to the period of the bit stream, whereas for RZ, the pulse occupies only a fraction of the period (typically, half the period). Figure 8.18 shows the bit sequence 10101101 in both the NRZ and RZ formats. In NRZ, when two consecutive 1's are present, the signal does not return to zero—hence the name. The major difference between NRZ and RZ pulse sequences is in the bandwidth requirement. Since for an RZ sequence the pulses are shorter and the rise and fall of the pulse intensity are faster, an RZ pulse sequence requires much higher bandwidth. Since the NRZ pulse format requires less bandwidth, the more commonly used format is NRZ, while for higher bit rates such as 40 Gb/s and higher, the RZ format is preferred, due to its greater immunity to nonlinear effects.

8.5 BIT ERROR RATE

The performance of digital communication systems is measured in terms of the *bit error rate* (BER), the rate at which errors are being committed by a receiver in correctly recognizing 1's and 0's. Thus, if in receiving n pulses (bits) consisting of a random sequence of 1's and 0's the receiver commits an average of r errors, the bit error rate is defined as the ratio of r to n. As an example, let us assume that in receiving 10 billion pulses the receiver commits five errors on average; the BER is 5×10^{-10}. If this is the BER for a 2.5-Gb/s communication system, then since in 1 second there would be 2.5×10^{9} bits, the average number of errors committed per second would be 1.25. If the system operates at 10 Gb/s with the same BER,

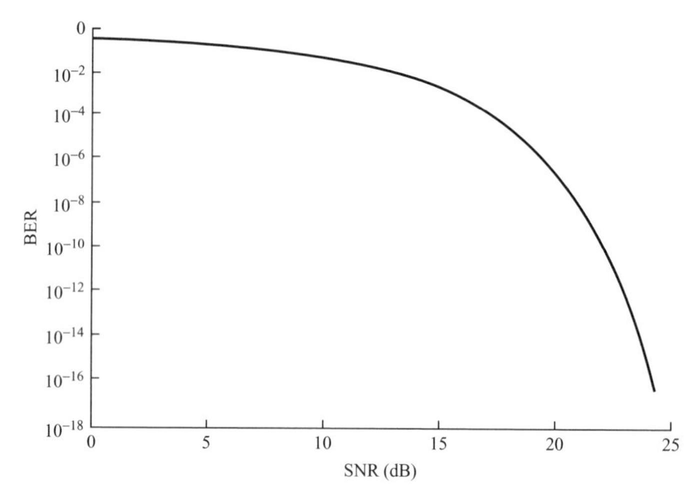

FIGURE 8.19 Variation of BER with SNR.

the number of errors per second would be 5. A typical BER requirement of current communication systems is 10^{-12} to 10^{-15}.

Since errors in detection can appear due to small incident optical powers and low signal-to-noise ratios, to achieve a BER smaller than a specified value, the photodetector needs a minimum received power, which is also dependent on the bit rate of the system. It is obvious that the higher the signal-to-noise ratio the lower would be the BER. For most PIN detectors the noise is dominated by thermal noise, which is independent of the signal power. Thus, the noise is the same in bits 1 and 0, and in such a case the optimum setting of a threshold value for the detection of pulses would be the midpoint of 1 and 0. In such a case we have an approximate expression for the relationship between BER and SNR which is valid for SNR > 72:

$$\text{BER} \approx \left(\frac{2}{\pi \cdot \text{SNR}} \right)^{1/2} e^{-\text{SNR}/8} \tag{8.6}$$

Using this equation it is easy to see that to achieve a BER of 10^{-9}, the SNR required is 144, or 21.6 dB on the decibel scale. Figure 8.19 shows the variation of BER with SNR. Note that the curve dips sharply beyond an SNR of 15 dB, and thus one can achieve a large improvement in detection even with small incremental improvements in SNR beyond 15 dB.

8.6 EYE DIAGRAM

One of the interesting ways to determine the quality of transmission is to use an *eye diagram*, in which the electrical signal generated by the photodetector is fed into an oscilloscope (an instrument that displays the electrical signal on a cathode-ray tube

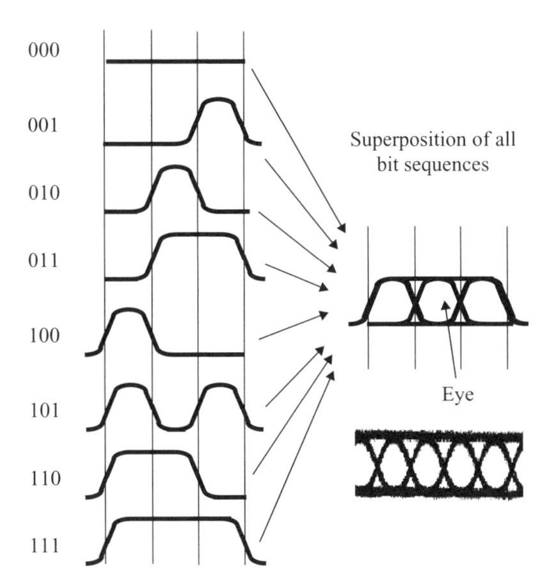

FIGURE 8.20 An eye pattern is formed by overlapping the random pulse sequence. In this case the eye is open.

such as that used in a television set). The oscilloscope is synchronized so that the pulses overlap on the screen. Since the bit pattern is random, if we consider a 3-bit pattern, the combinations can be only the following:

000, 001, 010, 011, 100, 101, 110, 111

Since 0 represents no light pulse and 1 represents the presence of a light pulse, at the transmitter side, assuming well-formed pulses, an overlap of all the pulses would look as shown in Fig. 8.20. The figure formed by the overlapping pulses is the eye diagram, and as can be seen, the center, called the *eye*, is open. An open eye implies that the pulses are easily resolvable. Figure 8.21 shows the eye pattern when the pulses have suffered dispersion and have some jitter (i.e., the time interval between adjacent pulses does not remain constant but keeps changing randomly). In this case the eye is not open, and detecting such pulses would result in significant bit error rates. An eye diagram is a very convenient method of visualizing the quality of pulses. A quick estimate of system performance can be obtained from the opening of the eye.

Figure 8.22 shows simulated eye patterns after signal propagation through 50 km of single-mode fiber for a directly modulated laser and an externally modulated laser. Since direct modulation introduces chirp in the laser output, its spectrum is broader, leading to enhanced dispersion. The external modulator, on the other hand, can give unchirped output pulses. The figure clearly shows the better eye pattern for the external modulator. Figure 8.23 shows a typical eye pattern as received after a signal has propagated through 500 and 9500 km of fiber, clearly showing eye degradation.

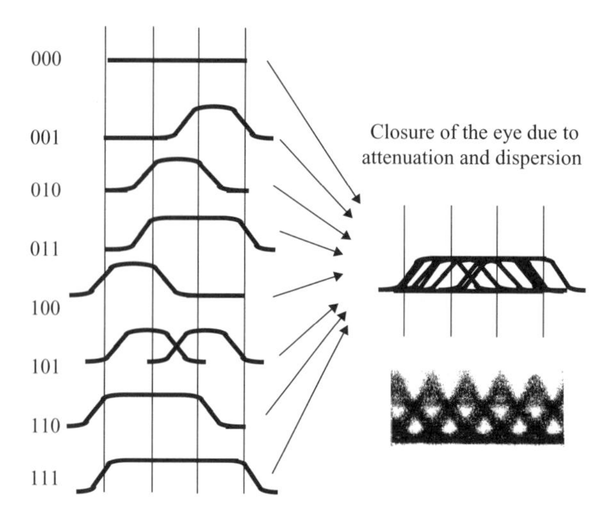

FIGURE 8.21 Closure of the eye due to attenuation and dispersion of pulses.

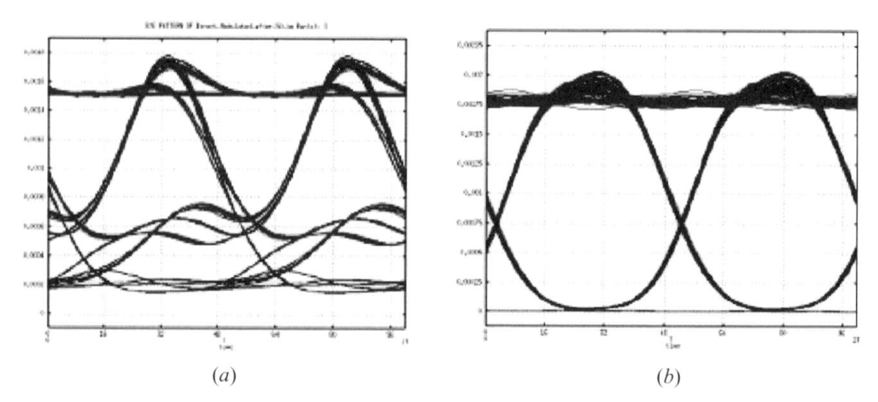

FIGURE 8.22 Simulated eye patterns after propagation through 50 km of single-mode fiber for (*a*) a directly modulated laser diode and (*b*) an externally modulated laser diode.

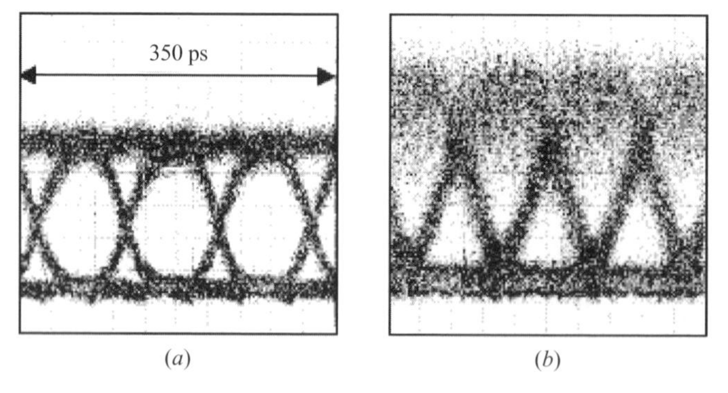

FIGURE 8.23 Eye patterns corresponding to propagation through (*a*) 500 and (*b*) 9500 km of fiber. Note the deterioration of the eye pattern at longer propagation distances. (After McGhan et al., 2005.)

8.7 LIMITATIONS DUE TO ATTENUATION AND DISPERSION

In the simplest fiber optic communication system, the point-to-point link, a transmitter at one end sends information along an optical fiber to a receiver at the other end. During propagation through the link the optical pulses suffer attenuation and dispersion. Both these effects would finally determine the power received, and the overlap between the pulses would in turn determine the bit error rate of the pulses. In order that the BER be below a certain specified value (typically, 10^{-12}), there must be a minimum power level at the receiver, and the pulse dispersion suffered by the pulses must be below a certain specified value. Both of these factors depend on the bit rate of the communication system.

Attenuation-Limited Distance

Assuming that the transmitter launches a certain optical power into the fiber optic link, the losses in the fiber, as well as in the connectors and splices that they would encounter along the link, would determine the power level received. Each of these is characterized by a loss in terms of decibels, and as this is a logarithmic scale, the total loss of the link can be estimated by adding the decibel loss of each component and the loss due to propagation.

If the attenuation coefficient of the fiber is α dB/km and the length of the link is L km, the loss due to propagation through the fiber would be αL dB. If there are N_s splices and N_c connectors in the link, and if α_s and α_c are the losses in decibels of each splice and each connector, respectively, the loss due to splices and connectors would be $(N_s\alpha_s + N_c\alpha_c)$ dB. Thus, the total loss suffered by the pulses would be $(\alpha L + N_s\alpha_s + N_c\alpha_c)$ dB and the power received by the receiver would be $P_i - (\alpha L + N_s\alpha_s + N_c\alpha_c)$ dBm, where P_i is the power in dBm at the transmitter end of the link. If the minimum power required at the receiver for operating below a specified bit rate is known, it is easy to estimate the maximum permissible link length. At this distance, either an electronic regeneration or an optical amplification would need to be carried out.

Example 8.4 Let us consider a link with 2 mW (= 3 dBm) of power input at the transmitter end. Let us also assume that the minimum power required at the receiver is -32 dBm ($\approx 0.6\ \mu$W), corresponding to a system operating at 2.5 Gb/s and using an APD photodetector. Thus, the maximum allowed loss of the signal between the transmitter and the receiver is $3 - (-32) = 35$ dB. If there are six splices, each with a loss of 0.2 dB, and two connectors, each with a loss of 0.1 dB, the loss allowed within the fiber link would be $35 - 6 \times 0.2 - 2 \times 0.1 = 33.6$ dB. If we allow a margin of 3.6 dB to take care of any eventualities that may arise later during the life of the link, we can permit a loss of 30 dB in the fiber. If the average fiber loss is 0.3 dB/km, the maximum distance that the signals can propagate would be 100 km. Any propagation beyond this length would require placing a regenerator or an optical amplifier at this point.

Power budgeting tells us the maximum distance that the signals can propagate, keeping in view the signal power that needs to be received by the detector.

Dispersion-Limited Distance

As discussed in Chapter 7, pulses suffer from broadening as they propagate through the fiber. If the broadening is large, adjacent pulses can start to overlap, resulting in loss of resolution of the bits and hence increased bit error rate. In the case of multimode fibers, if the total dispersion is given by $\Delta\tau$, then using NRZ coding, the maximum permissible dispersion limited bit rate is given approximately by

$$B_{\max} \approx \frac{0.7}{\Delta\tau} \tag{8.7}$$

Operation around 1310 nm minimizes the material dispersion, and hence almost all multimode fiber systems operate in this wavelength region, with optimum refractive index profiles having small values of intermodal dispersion.

Let us consider a communication system that needs to operate at 2.5 Gb/s. Using Eq. (8.7), we see that the maximum allowable pulse dispersion is about 280 ps. Assuming that the source and detector have sufficient bandwidth, the maximum dispersion of the fiber should be 280 ps. If we assume an LED operating at 1310 nm with a spectral width of 25 nm, the material dispersion contribution would be negligible compared with the intermodal dispersion, since silica fibers have almost zero material dispersion around 1300 nm. If the fiber is a parabolic-index fiber with $\Delta = 0.01$, assuming that $n_1 \approx 1.46$, we have an intermodal dispersion of 240 ps/km [see Eq. (6.12)]. Hence, with this fiber, the maximum distance that the signals can propagate is about 850 m. In case the transmitter and receiver have comparable rise times, the maximum allowed distance would be reduced further. At other operating wavelengths, there would be additional dispersion due to material contribution and the maximum allowed distance would be reduced even further.

In the case of single-mode fibers, the source is usually a very narrow spectral width source, and when such sources are pulsed, the spectral width of the resulting pulse is given approximately by the inverse of the pulse duration. Thus, shorter pulses would have larger spectral width, and wider pulses would have narrower spectral widths. In such a case, we can write the following relationship connecting the maximum bit rate B for a given link length L and dispersion coefficient D operating at 1550 nm:

$$B^2 DL < 1.9 \times 10^5 \text{ Gb}^2/\text{s} \cdot \text{nm} \tag{8.8}$$

where B is measured in Gb/s, D in ps/km·nm and L in km.

Example 8.5 Consider a system that is required to operate at 2.5 Gb/s. At this bit rate, the maximum allowed dispersion (DL) is approximately 30,400 ps/nm. Hence if the fiber has a dispersion of 16 ps/km·nm (G.652 standard single-mode fiber with zero dispersion at 1310 nm operating at 1550 nm), the maximum dispersion-limited distance would be about 1900 km. If the fiber is a nonzero-dispersion-shifted fiber (G.655 fiber) with a dispersion coefficient of 3 ps/km·nm, the maximum allowed distance would be about 10,000 km. If the G.655 fiber is to operate at 10 Gb/s,

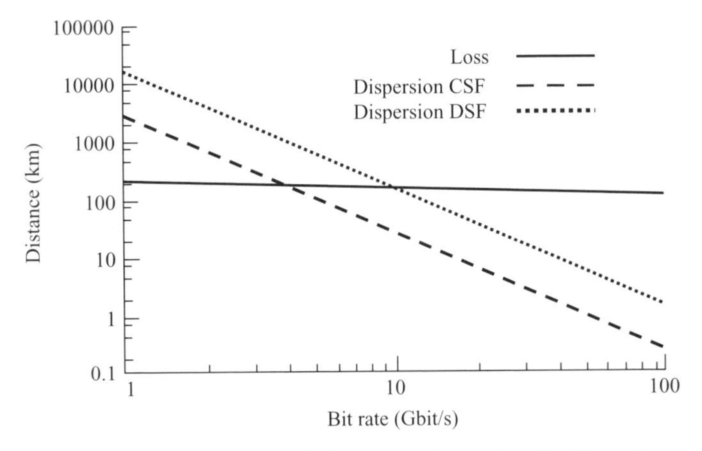

FIGURE 8.24 Maximum unrepeatered distance as determined by dispersion and attenuation in the case of conventional single-mode fiber and dispersion-shifted fiber.

the maximum allowed dispersion would be 1900 ps/nm and the maximum distance would be about 600 km.

Figure 8.24 shows the maximum unrepeatered length as imposed by pulse dispersion in conventional single-mode fibers and dispersion-shifted fibers operating at 1550 nm, and also the limit imposed by attenuation. These limits have been calculated assuming that the spectral width is determined by the pulse width [Eq. (8.8)]. At lower bit rates the maximum unrepeatered distance limits are due primarily to attenuation, while at higher bit rates, the maximum distance is limited by dispersion. Both these limits can be overcome by the use of optical amplifiers and dispersion compensators.

Figure 8.25 gives the rate at which the achievable bit rates have been increasing over a period of time. It is apparent that the bit rates have increased by about a factor of 6 in the past five years. Apart from the increase in bit rates, there is a continuous increase in the number of wavelength channels with decreased wavelength spacing between adjacent channels. Of course, the separation between adjacent channels

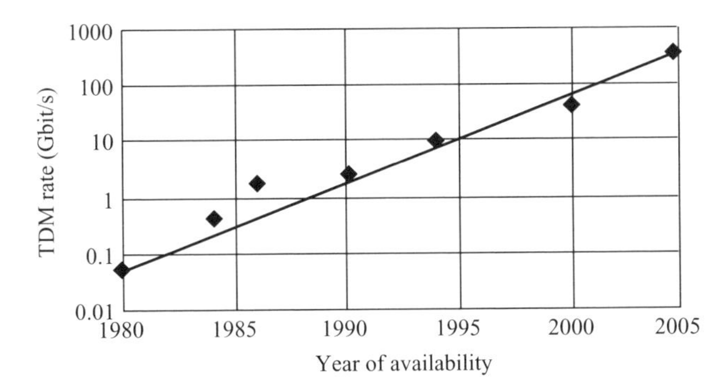

FIGURE 8.25 Bit rates have increased by a factor of 6 within the last five years. (After B. Mikkelsen, G. Raybon and R J Essiambre, Ultra high-data-rate optical communication.)

cannot be less than the bandwidth occupied by the signal itself. Thus, a 10-Gb/s system occupies a bandwidth of about 10 GHz, and the channel spacing to the adjacent channel has to be larger than this, which corresponds to a wavelength spacing of about 0.08 nm.

As we saw earlier, the performance of a fiber optic communication system is determined by attenuation, dispersion, and nonlinearity; the last aspect is discussed in detail in Chapter 13. The transmitter used in the link is characterized by its output power, the maximum rate at which it can be modulated internally in the case of direct modulation, and the dispersion that it can tolerate. Similarly, the receiver is characterized by the sensitivity and speed of response and determines the minimum power required to achieve a given bit error rate at the transmission speed. The optical fiber attenuates the signal, disperses it, and leads to random effects between bits or between channels, due to nonlinear effects. Optical amplifiers placed within the link lead to compensation of the loss, but as we shall see in Chapter 9, they also add to noise in the signal. Dispersion-compensating elements placed at various points along the link compensate for accumulated dispersion but also attenuate the signal. This attenuation has to be compensated by additional amplifiers placed along with the dispersion-compensating elements. The overall performance of a fiber optic system is determined by these effects, and optimization of the various elements and their placement within the link is an issue of design.

Figure 8.26 is a world map that shows the major undersea fiber optic systems installed up to 2005. With increased penetration of optical fibers, earth will have a fiber optic network connecting any place to any other place.

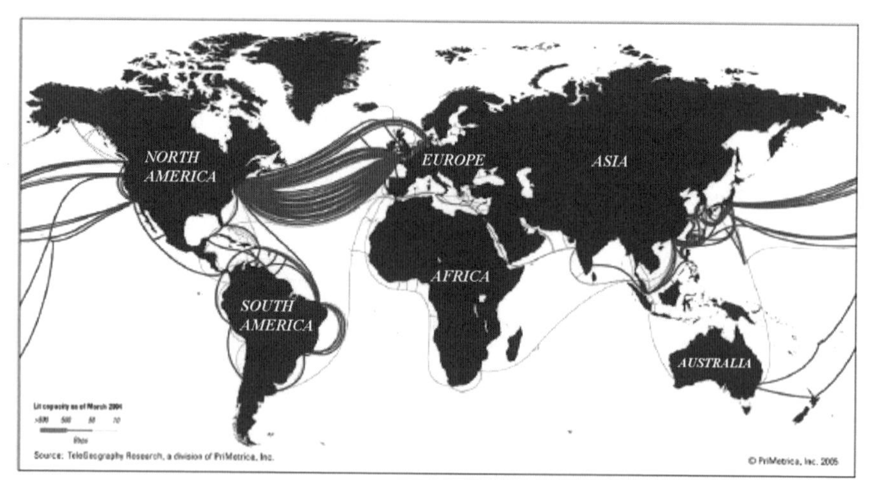

FIGURE 8.26 World map with major undersea fiber optic systems. (After Bergano, 2005. Copyright © 2005 IEEE.)

CHAPTER 9

Erbium-Doped Fiber Amplifiers and Fiber Lasers

9.1 INTRODUCTION

In traditional long-distance optical fiber communication systems, compensation of loss and dispersion is usually accomplished using electronic regenerators. In an electronic regenerator (Fig. 9.1), attenuated and temporally broadened optical pulses are first detected by a photodetector (shown as "PD" in the figure), which converts the optical signals to electrical signals. These are then processed, retimed, cleaned of noise, and amplified electronically, and the amplified electrical pulses drive a laser diode (shown as "LD") to produce optical pulses almost identical to the original. These regenerated pulses are then launched into the fiber link for transmission to the next regenerator. Such regenerators are designed to operate at a specific bit rate, and when there is a need to upgrade the system to higher bit rates, the regenerator needs to be replaced. Since each regenerator operates at one optical wavelength, when there is a requirement to increase the capacity by using another signal at another wavelength through the same fiber, then at the regenerator sites, the signals at different wavelengths first need to be demultiplexed (separated) into separate paths, and each wavelength then needs to be processed individually by separate regenerators. After regeneration of the individual signals at different wavelengths, the signals need to be multiplexed (combined) into a single output for further transmission through a fiber. Such regenerators not only compensate for the loss and dispersion suffered by the signal but also clean the signal of any accumulated noise, and thus the re-formed pulses after the regenerators are almost as good as at the beginning of their journey. At the same time, for wavelength-division-multiplexed transmission systems carrying multiple wavelength signals through one fiber, a solution using electronic regeneration would be very expensive.

Whenever the system limitation is due to insufficient optical power rather than dispersion, what is needed is simply amplification of the signal, and optical amplifiers can perform this job very well. Optical amplifiers are devices that amplify the

---

Fiber Optic Essentials, By K. Thyagarajan and Ajoy Ghatak

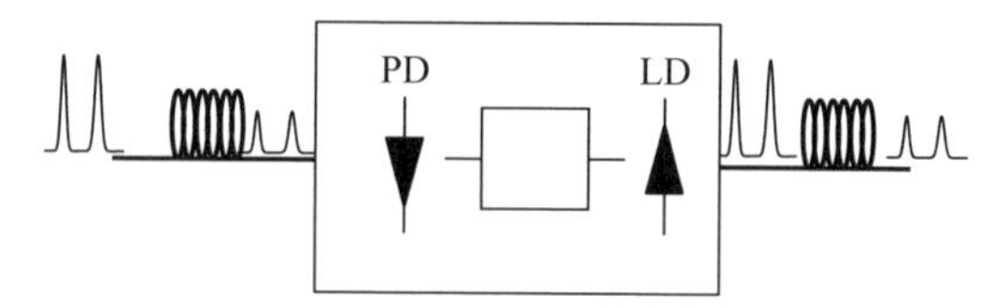

FIGURE 9.1 Electronic regenerator used to transmitt optical signals over long distances. When the optical power drops, the electronic regenerator converts the optical signals into electrical signals, processes them electronically, and then converts them back into optical signals using a laser diode. The signals coming out of the regenerator are almost as good as at the beginning of their journey. PD, photodetector; LD, laser diode.

incoming optical signals in the optical domain itself without conversion to the electrical domain, and have truly revolutionized long-distance fiber optic communications.

Optical amplifiers have two advantages over electronic regenerators: They do not need high-speed electronic circuitry, and, they are transparent to bit rate and format and most important, can amplify multiple optical signals at different wavelengths simultaneously. Their development has ushered in a tremendous growth in communication capacity using wavelength-division multiplexing (WDM), in which multiple wavelengths carrying independent signals are propagated through the same single-mode fiber, thus multiplying the capacity of the link (Fig. 8.1). Of course, compared to electronic regenerators, they also have drawbacks: They do not compensate for dispersion accumulated in the link, and they add noise to the optical signal. As we will see later, this noise leads to a maximum number of amplifiers that can be cascaded so that the signal-to-noise ratio is within the limits.

Optical amplifiers can be used at many points in a communication link. Figure 9.2 shows some typical examples. A booster amplifier is used to boost the power of the transmitter before launching into the fiber link. The increased transmitter power can be used to go farther in the link. The preamplifier placed just before the receiver is used to increase the receiver sensitivity (the minimum power required by the receiver to function properly). Inline amplifiers are used at intermediate points in the link to overcome fiber transmission and other losses. Optical amplifiers can also be used for overcoming splitter losses: for example, for distribution of cable television.

There are currently three principal types of optical amplifiers: the *erbium-doped fiber amplifier* (EDFA), the *Raman fiber amplifier* (RFA), and the *semiconductor optical amplifier* (SOA). Today, most optical fiber communication systems use

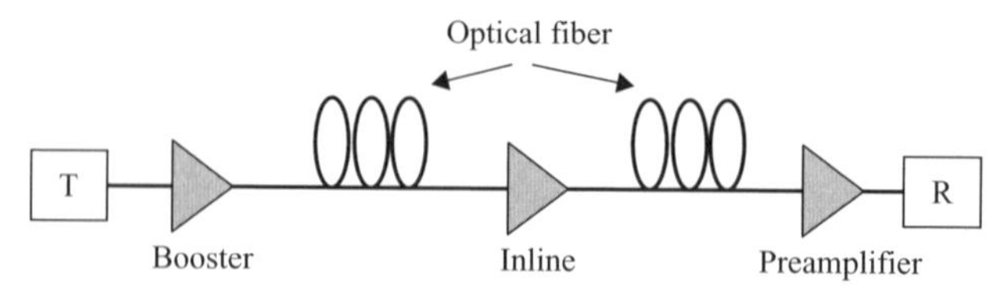

FIGURE 9.2 Typical fiber optic system with optical amplifiers as booster, inline, and preamplifiers.

EDFAs, due to their advantages in terms of bandwidth, high power output, and noise characteristics. RFAs and SOAs are also becoming important in many applications. In the following sections we discuss the characteristics of EDFAs, and in Chapter 10 we discuss RFAs. Detailed account of EDFAs can be found in Desurvire (1994).

9.2 PRINCIPLES OF THE ERBIUM-DOPED FIBER AMPLIFIER

Optical amplification by EDFA is based on the process of stimulated emission, which is the basic principle behind laser operation. In fact, a laser without optical feedback is just an optical amplifier.

Atoms and molecules are characterized by discrete energy levels and they make transitions between these levels of energy whenever they absorb or emit electromagnetic radiation. The yellow color emission from sodium lamp is due to emission by sodium atoms when they jump from a higher-energy state to a lower-energy state. The emission wavelength is characteristic of the atom that is emitting the light. Thus, neon lights are red in color, due to the energy levels of neon atoms. Figure 9.3 shows two lowest-lying energy levels of an atomic system: the ground level with energy E_1 and an excited level with energy E_2. These are the lowest-lying energy states that the atom or molecule can occupy. The atom or molecule cannot have an intermediate energy value between the two energy values shown. Thus, we can say that the energy of the atom or molecule is quantized.

The atom described by different energy levels can interact with electromagnetic radiation in three distinct ways: absorption, spontaneous emission, and stimulated emission.

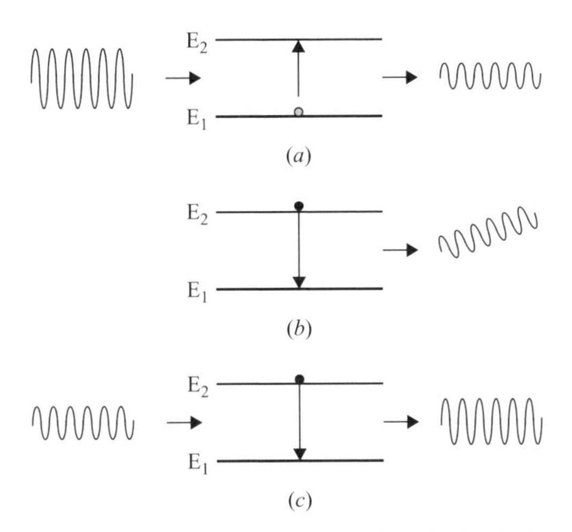

FIGURE 9.3 Atoms can interact with electromagnetic radiation in three distinct ways: (*a*) absorption; (*b*) spontaneous emission; (*c*) stimulated emission.

1. ***Absorption.*** In the case of absorption, an atom occupying a lower-energy state can absorb radiation of an appropriate wavelength and get excited to an upper energy level (Fig. 9.3*a*). The atom occupying energy level E_1 can absorb radiation at a frequency ν_0 given by the following equation and get excited to the level with energy E_2:

$$\nu_0 = \frac{E_2 - E_1}{h} \tag{9.1}$$

   where h, Planck's constant, has a value of 6.634×10^{-34} J·s. Since the energy values of the various levels are dependent on the atom, an atom will absorb light of certain wavelengths only, which correspond to the various pairs of energy levels. For example, corresponding to a wavelength of about 589 nm in the yellow region of the visible spectrum, the frequency will be given by

$$\nu = \frac{c}{\lambda} = \frac{3 \times 10^8}{589 \times 10^{-9}} \approx 5.094 \times 10^{14}\ \text{Hz}$$

   Thus,

$$E_2 - E_1 = h\nu \approx 3.39 \times 10^{-19}\ \text{J}$$

   To give an idea of the magnitude of this energy value, note that when we lift a 1-kg weight by 1 m, we perform about 10 J of work, and this energy is stored in the form of potential energy in the mass. If you drop this weight from this height, the energy can be used to do work. So 3.4×10^{-19} J is really a very, very small energy value!
2. ***Spontaneous emission.*** An atom occupying an upper level can radiate electromagnetic radiation spontaneously and deexcite itself to the lower level (Fig. 9.3*b*), a phenomenon known as *spontaneous emission*. If the energy levels are the same as considered above, the frequency of the radiation emitted will again be ν_0. Spontaneous emissions are completely random and appear in all directions. Light coming from most optical sources, including the sun, is due primarily to spontaneous emission.
3. ***Stimulated emission.*** Apart from these two processes, an atom occupying the upper energy level can also be stimulated to emit radiation at the frequency ν_0 by an incident light wave at that frequency (Fig. 9.3*c*) in a process called *stimulated emission*. The primary difference between spontaneous and stimulated emission is that whereas the former emission is completely random in direction, polarization, and so on, the latter is coherent with the incident radiation. This implies that the radiation emitted by the atom is identical in all respects to the radiation that stimulates the atom, and in this process the incident radiation gets coherently amplified by the stimulated emission process.

We may mention here that in an emission process the radiation is not monochromatic but is spread over a certain frequency range. Thus, energy levels have a certain width (usually referred to as *line width*), and atoms can interact over a range of frequencies. As an example, we consider the emission from sodium lamps. Sodium lamps are used worldwide for illumination, including street lighting. When an electric discharge is passed through a sodium lamp, which consists of a discharge tube filled with sodium gas, the electrons speeding up in the tube from the negative terminal to the positive terminal bombard the sodium atoms and excite them from the ground state to an excited state. Upon reaching the excited state, the sodium atoms spontaneously emit light corresponding to the energy difference between the energy levels. This energy difference corresponds to light yellow in color, and hence the sodium lamp glows yellow. The tube also contains a bit of neon and argon to start the gas discharge, which is why when the lamp is switched on, it initially emits a pinkish or reddish light.

The concept of stimulated emission was first put forth by Albert Einstein in 1917 and forms the basis of lasers, which are finding applications in all branches of science, technology, and engineering. It so happens that for a given atomic system and for a given pair of energy levels of the atom, the three processes (absorption, spontaneous emission, and stimulated emission) are related to each other, and in particular, the probability of absorption is the same as the probability of stimulated emission; this was predicted by Einstein and later confirmed by quantum mechanics.

Now, when the atomic system is in thermal equilibrium (i.e., in equilibrium with the surroundings), most of the atoms will be found in the ground level. Thus, if light at a specific wavelength (corresponding to the atom) falls on this collection of atoms, it will result in a greater number of absorptions (from ground level to upper level) than stimulated emissions (from upper level to ground level), and the light beam will suffer from attenuation (Fig. 9.4). On the other hand, if the number of atoms in the

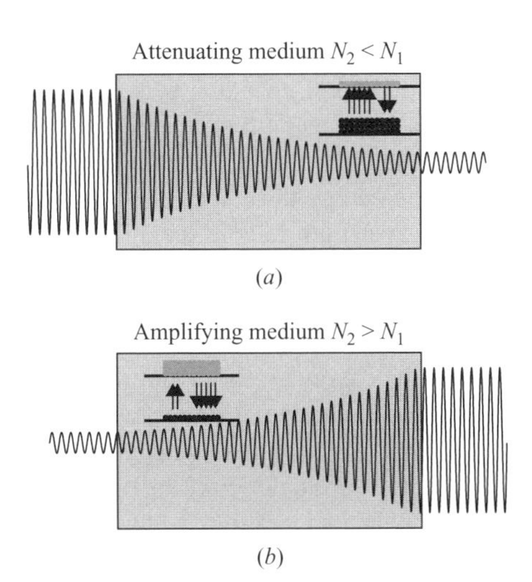

FIGURE 9.4 (*a*) Under normal equilibrium conditions, there are more atoms in the ground state than in an excited state, and an incident light wave undergoes attenuation. (*b*) When there is population inversion, the light beam gets amplified, due to the process of stimulated emission.

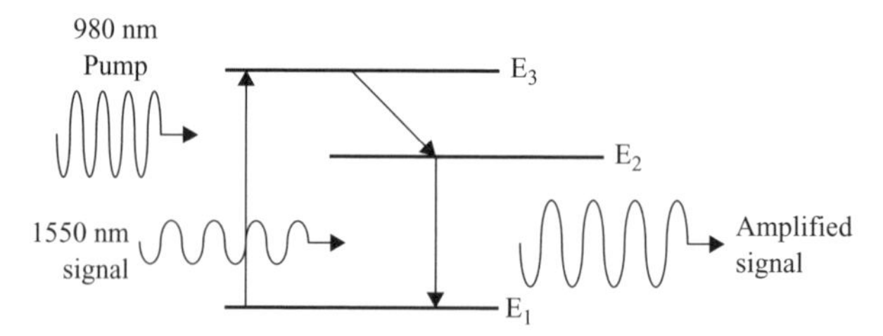

FIGURE 9.5 Three lowest-lying energy bands of erbium in silica matrix. The pump laser at 980 nm excites the erbium ions from the ground level E_1 to the level marked E_3, from which they make a nonradiative transition to level E_2. Level E_2 is a metastable level, and population inversion between levels E_2 and E_1 is responsible for the amplification of signals in the 1550-nm band.

upper level could be made greater than those in the lower level, an incident light beam at the appropriate wavelength could induce more stimulated emissions than absorptions, thus leading to optical amplification. Known as *light amplification by stimulated emission*, this is the basic principle behind an EDFA.

Figure 9.5 shows the three lowest-lying energy levels of erbium ion located within silica glass. Light from a semiconductor laser at 980 nm (called a *pump laser*) excites erbium ions from the ground state to the level marked E_3 i.e., erbium atoms in the ground state absorb the 980-nm radiation and get excited to the level marked E_3. We may mention here that the photons corresponding to the 980-nm wavelength have an energy of about 2×10^{-19} J, which represents the energy difference $E_3 - E_1$. Level E_3 is a short-lived energy level; After a few microseconds, ions from this level jump down to level E_2. The lifetime of level E_2 is much longer, about 12 ms. Hence, ions brought to level E_2 stay there for a significantly longer time. Thus, by pumping hard enough, the population of ions in level E_2 can be made larger than the population of level E_1 thereby achieving population inversion between levels E_1 and E_2. In such a situation, if a light beam at a wavelength corresponding to the energy difference $(E_2 - E_1)$ falls on the collection, it will get amplified by the process of stimulated emission. For erbium ions, the energy difference $E_2 - E_1$ is approximately 1.28×10^{-19} J, the corresponding wavelength falls in the 1550-nm band, and thus it is an ideal amplifier for signals in the 1550-nm window. The process discussed above is usually referred to as *fluorescence*.

Now, in the case of erbium ions located within silica glass, due to interactions between neighboring atoms, the energy levels are not sharp levels but are broadened: that is, ions can have energies over a range of values, which implies that as they jump from the higher level to the lower level, their wavelengths can have a range of values. Hence the system is capable of absorbing or emitting over a band of wavelengths and consequently, of amplifying optical signals over a band of wavelengths.

Figure 9.6 is a schematic of an EDFA that consists of a short piece (about 20 m in length) of erbium-doped fiber (EDF), a single-mode fiber doped with erbium (typically, with 100 to 500 parts per million) in the core, and which is pumped by a 980-nm pump laser through a wavelength-division-multiplexing (WDM) coupler.

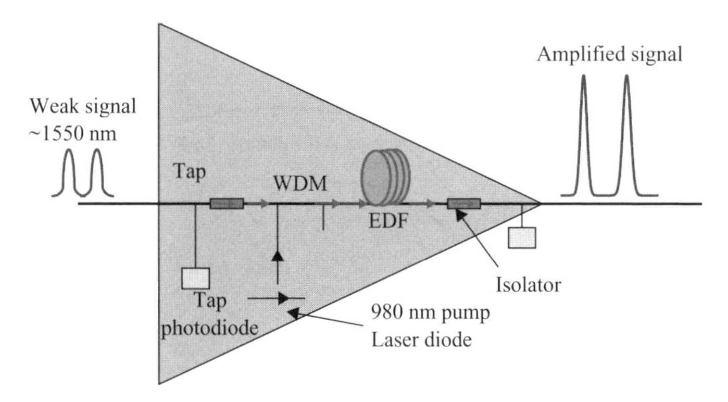

FIGURE 9.6 Schematic of an EDFA consisting of a 980-nm pump laser, WDM coupler, and short piece of erbium-doped fiber. Tap couplers are used to monitor the input and output from the amplifier, and the isolator prevents reflected light from entering the EDFA.

The WDM coupler (see Chapter 12) multiplexes (combines) light of wavelengths 980 and 1550 nm from two different input fibers to a single output fiber. The 980-nm pump light is absorbed by the erbium ions to create population inversion between levels E_2 and E_1. Thus, incoming signals in the 1550-nm wavelength region get amplified as they propagate through the population-inverted doped fiber. The tap couplers are couplers that tap a very small fraction of the light from the input and output to make it possible to measure the signal power entering and exiting an amplifier. These values are used to control the amplifier for constant gain or constant ouput power operation. The isolator is a device that allows light to propagate along only one direction. The isolator is placed to prevent any reflected light from entering the amplifier, which otherwise can get destabilized and start to oscillate like a laser.

There are three important parameters of any optical amplifier: gain and gain spectrum, saturation behavior, and noise. We discuss these in the following sections.

9.3 GAIN AND GAIN SPECTRUM

The *gain* of an EDFA is defined as the ratio of the output signal power (P_{out}) to the input signal power (P_{in}):

$$\tilde{G}(\text{dB}) = 10 \log \frac{P_{\text{out}}}{P_{\text{in}}} \tag{9.2}$$

The gain depends on the doping concentration and doping profile of the erbium-doped fiber, the length of the fiber, and the pump power. Typical gain values of an EDFA are about 20 to 30 dB i.e., the output power is about 100 or 1000 times the input power. The gain provided by the amplifier depends on the erbium doping in the doped fiber, the length of the fiber, and the pump power. These parameters are usually optimized for achieving the desired gain characteristics.

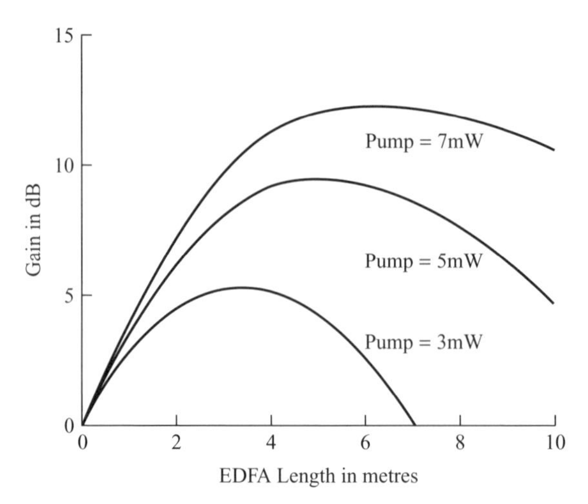

FIGURE 9.7 Variation of gain with EDFA length for different values of pump powers. For a given pump power there is an optimum length for achieving maximum gain.

For a given input pump power, as we increase the length of the doped fiber, the gain would first increase and then after reaching a maximum would start to decrease. This happens because as the pump propagates through the doped fiber it gets absorbed and thus its power reduces. After propagating a certain distance, its power is too small to create population inversion, and thus after this length, the doped fiber would start to absorb the signal rather than amplify it. Figure 9.7 shows the variation of gain with the length of the doped fiber for different pump powers. Hence, for a given pump power there is an optimum length of the doped fiber to achieve maximum gain.

For a given length of the doped fiber, as the pump power increases, we expect the gain to increase. At the same time, as the pump power increases it creates more and more population inversion, and once all erbium ions in the fiber are excited, no more erbium ions are available and hence the gain would saturate. Figure 9.8 shows a typical variation of gain with input pump power for different lengths of the doped fiber, clearly showing gain saturation with increase in pump power.

Figure 9.9 shows typical measured gain spectra of an EDFA for low input signal power (-25 dBm $= 3.16\ \mu$W) and for higher input signal power (-5 dBm $= 0.3$ mW). As can be seen, EDFAs can provide amplifications of greater than 20 dB (power amplification by a factor greater than 100) over the entire band of 40 nm from 1525 to about 1565 nm. This wavelength band, referred to as the *C-band* (conventional band), is the most common wavelength band of operation. With proper amplifier optimization, EDFAs can also amplify signals in the wavelength range 1570 to 1610 nm; this band of wavelengths is referred to as the *L-band* (long-wavelength band). C- and L-band amplifiers can be used simultaneously to amplify 160 different wavelength channels. Such systems are now available commercially.

It can be seen from Fig. 9.9 that although EDFAs can provide gains over an entire band of 40 nm, the gain is not flat (i.e., the gain depends on the signal wavelength).

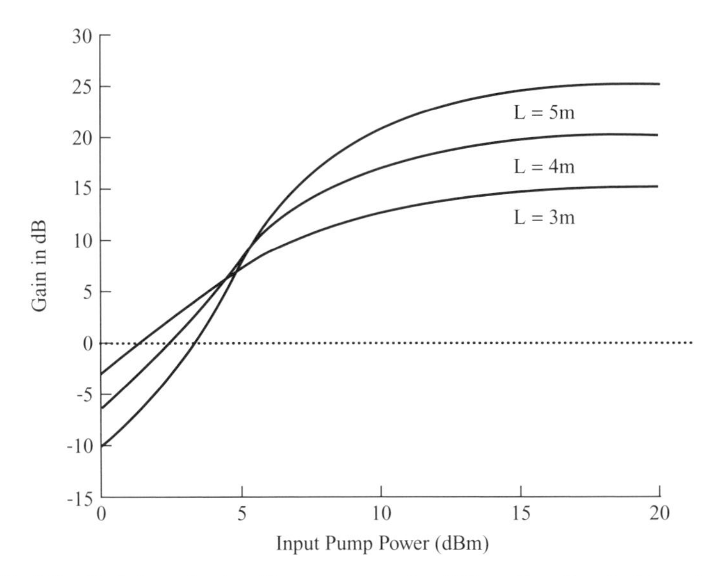

FIGURE 9.8 Variation of gain with pump power for different lengths of erbium-doped fiber.

This is especially true for small-input signal powers. Thus, if multiple wavelength signals with the same power are input to the amplifier, their output powers will differ. In a communication system employing a chain of amplifiers, differential signal gain among the various signal wavelengths (channels) from each amplifier will result in a significant difference in signal power levels and hence in the signal-to-noise ratio (SNR) among the various channels. In fact, signals for which the gain in the amplifier is greater than the loss suffered in the link will keep on increasing in power level, while channels for which the amplifier gain is less than the loss suffered will continue

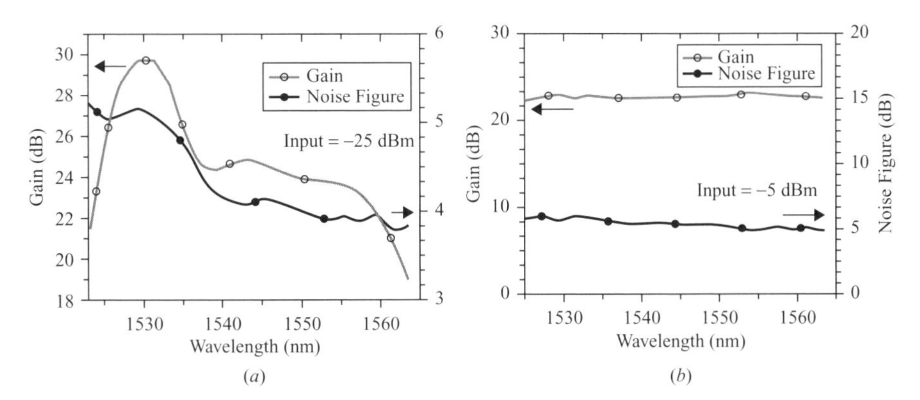

FIGURE 9.9 (*a*) Small-signal gain and noise figure of EDFA for an input signal power of −25 dBm; (*b*) gain and noise figure at input power −5 dBm.

to decrease in power. The former channels will finally saturate the amplifiers and will lead to increased nonlinear effects in the link, while the latter will have reduced SNR, leading to increased errors in detection. Thus, such a differential amplifier gain is not desirable in a communication system, and it is very important to have gain-flattened amplifiers.

Example 9.1 Consider an 80-km-long optical fiber having an attenuation of 0.28 dB/km at 1550 nm. Let us calculate the amplifier gain required to compensate for the loss in this link. Now,

$$\text{loss over } 80\,\text{km} = 80 \times 0.28 = 22.4\,\text{dB}$$

Hence the gain of the amplifier should be 22.4 dB. An input signal with a power of 1 mW (= 0 dBm) would have a power of −22.4 dBm (= 5.75 μW) at the exit of the fiber, which would be the input power to the amplifier.

Example 9.2 Consider an EDFA with a gain of 20 dB. If the input signal power is 1 μW (= −30 dBm), since a gain of 20 dB corresponds to a factor of 100, the output power will be 100 μW (= 0.1 mW). In dBm units, the input power of 1 μW corresponds to −30 dBm. For a gain of 20 dB, the output would be (–30 + 20) dBm = −10 dBm, which corresponds to a power of 0.1 mW. Note that multiplication (division) in linear units become addition (subtraction) in logarithmic, dB, and dBm units.

Example 9.3 From Fig. 9.7 it can be seen that the gain at 1530 nm is 35 dB, while that at 1535 nm is 30 dB. The output powers for an input power of 1 μW at 1530 and 1535 nm are

$$\text{output power} \begin{cases} -30\,\text{dBm} + 35\,\text{dB} = +5\,\text{dBm} = 3.2\,\text{mW} \\ -30\,\text{dBm} + 30\,\text{dB} = 0\,\text{dBm} = 1\,\text{mW} \end{cases}$$

which are significantly different, due to different gains at the two wavelengths.

9.4 GAIN FLATTENING OF EDFAs

There are basically two main techniques for gain flattening: One uses external wavelength filters to flatten the gain while the other one relies on modifying the amplifying fiber properties to flatten the gain. In gain flattening using external filters, the output of the amplifier is passed through a special wavelength filter whose transmission characteristic is exactly the inverse of the gain spectrum of the amplifier (see Fig. 9.10). Thus, channels that have experienced greater gain in the amplifier will suffer greater transmission loss as they propagate through the filter, while channels that experience smaller gain will suffer a smaller loss. By tailoring the filter transmission profile appropriately, it is possible to flatten the gain spectrum of the amplifier.

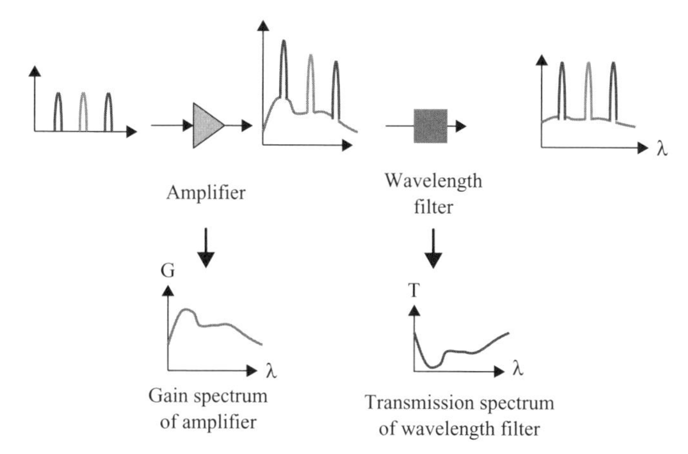

FIGURE 9.10 Principle of gain flattening in EDFAs. The filter transmission profile is exactly opposite the gain profile of the amplifier, resulting in gain flattening.

Placing the gain-flattening filter after the amplifier will result in reduction of the net gain of the amplifier. On the other hand, if the filter is placed prior to the signal entering the amplifier, one finds that this results in increased amplifier noise. Thus, in practical amplifiers, the gain-flattening filter is usually placed within the amplifier (i.e., the filter is placed after a certain length of the doped fiber, and the filter is followed by another piece of doped fiber). In this way, one can optimize the amplifier for maximum gain and reduced noise while retaining a flat gain spectrum.

Filters with specific transmission profiles can be designed and fabricated using various techniques. These include thin-film interference filters and filters based on short- or long-period fiber gratings (LPGs). Typical gain flatness of better than 1 dB can be achieved, and gain-flattened EDFAs are available commercially (Fig. 9.11).

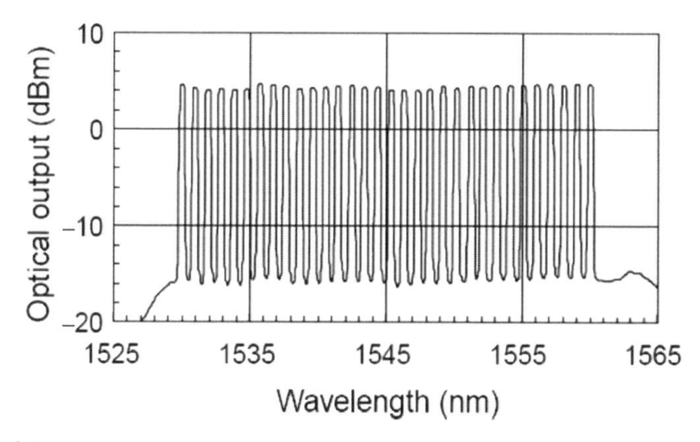

FIGURE 9.11 Output from a gain-flattened amplifier wherein 32 WDM channels are incident at the input of the amplifier. (After Mizuno et al., 2000.)

The gain variation over the entire C-band can be reduced to below 0.5 dB. In Chapter 10 we discuss fiber gratings and gain flattening using long-period gratings.

The wavelength variation of gain of an EDFA depends on the fiber properties as well as the doping level and where the erbium ions are actually located within the fiber. It is possible to achieve significant gain flattening by an appropriate choice of fiber type and also by controlling the doping of erbium in the fiber. EDFAs realized using such fibers can lead to lower cost since the number of components used in the amplifier decreases. Such a technique is referred to as *intrinsic gain flattening*.

Example 9.4 Let us consider an amplifier with a gain difference of 1 dB between two signal wavelengths. If the actual gains are 20 and 21 dB at these two wavelengths, let us estimate the difference in output powers for input powers of 10 μW each. For the wavelength with 20-dB gain, the output would be 1 mW, while that for the wavelength with 21-dB gain would be about 1.26 mW.

Example 9.5 Let us consider a long link with eight amplifiers operating at two wavelengths λ_1 and λ_2 with gains of 20 and 21 dB, respectively. Let us assume that the signal powers as they enter the first span are 1 mW each. Let us also assume that the span loss is 20 dB at both wavelengths. Now, for λ_1, the output signal power after eight amplifiers will be 1 mW since the gain exactly compensates for the loss. On the other hand, for λ_2, the loss in each span is only 20 dB while the gain is 21 dB. Hence, the corresponding signal powers at the end of eight spans would be 6.3 mW, since in each span the signal gains 1 dB of power (loss of 20 dB and a gain of 21 dB). This shows the large signal power differential that can exist at the end of a link due to a small gain variation in each amplifier.

9.5 NOISE IN EDFA

If EDFAs can compensate for the loss suffered while propagating through a fiber, the question that arises in one's mind is whether it is possible to traverse an arbitrarily long distance in the fiber by periodic amplification along the fiber link provided that the dispersion effects do not limit the distance. This is, in fact, not possible, due to the addition of noise by each amplifier, as discussed below.

In an EDFA, population inversion between two energy levels of erbium ion leads to optical amplification by the process of stimulated emission. As mentioned earlier, erbium ions occupying the upper energy level can also make spontaneous transitions to the ground state and emit radiation. This radiation appears over the entire fluorescent band of emission of erbium ions and travels in both the forward and backward directions along the fiber. Just like the signal, the spontaneous emission generated at any point along the fiber can be amplified as it propagates through the population-inverted fiber. The resulting radiation is called *amplified spontaneous emission* (ASE). This ASE, which has no relationship with the signal propagating through the amplifier, is the basic mechanism leading to noise in the optical amplifier.

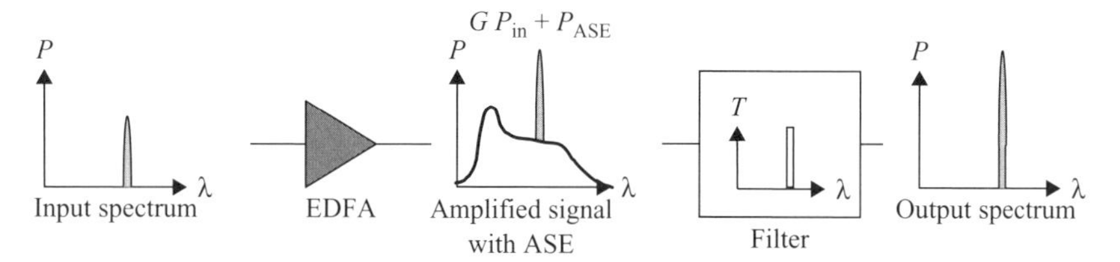

FIGURE 9.12 An EDFA amplifies an input signal, and along with the amplified signal there is a background ASE that constitutes the noise of the amplifier. Any ASE not coincident with the signal wavelength can be filtered using an optical filter. However, ASE within the signal band cannot be filtered and constitutes the minimum noise added by the amplifier.

Figure 9.12 shows the spectrum at the input of an EDFA and the output from the EDFA. At the output we have both the amplified signal and a background ASE. The ASE appearing in a wavelength region not coincident with the signal can be filtered using an optical filter as shown in the figure. On the other hand, the ASE that appears in the signal wavelength region cannot be separated and constitutes the minimum added noise from the amplifier.

If P_{in} represents the signal input power (at frequency ν) into the amplifier and G represents the gain of the amplifier in linear units (the corresponding gain in decibels is given by $\tilde{G} = 10 \log G$), the output signal power is given by GP_{in}. Along with this amplified signal, there is ASE power, which can be shown to be given by (see, e.g., Becker et al., 1999)

$$P_{\mathrm{ASE}} \approx 2n_{\mathrm{SP}}(G - 1)h\nu B_o \tag{9.3}$$

where B_o is the optical bandwidth in the frequency domain over which the ASE power is being measured (which must be at least equal to the optical bandwidth of the signal), and the spontaneous emission factor n_{sp} is given by

$$n_{\mathrm{SP}} = \frac{N_2}{N_2 - N_1} \tag{9.4}$$

Here N_2 and N_1 represent the population densities (number of atoms per unit volume) in the upper and lower amplifier energy levels of erbium in the fiber. Minimum value for n_{sp} corresponds to a completely inverted amplifier for which $N_1 = 0$ (i.e., all atoms excited to the upper level) and thus $n_{\mathrm{sp}} = 1$; for partial inversion, $n_{\mathrm{sp}} > 1$.

As a typical example, we have $n_{\mathrm{sp}} = 2$, $G = 100$ ($\tilde{G} = 20\,\mathrm{dB}$), $\lambda = 1550$ nm, and $B_o = 12.5$ GHz ($= 0.1$ nm at 1550 nm), which gives

$$P_{\mathrm{ASE}} = 0.6\,\mu\mathrm{W}(= -32\,\mathrm{dBm})$$

which corresponds to an ASE noise spectral density (noise power per unit bandwidth) of -22 dBm/nm. We can define the *optical signal-to-noise ratio* (OSNR) as the ratio of the output optical signal power to the ASE power:

$$\text{OSNR} = \frac{P_{\text{out}}}{P_{\text{ASE}}} = \frac{G P_{\text{in}}}{2n_{\text{sp}}(G-1)h\nu B_o} \tag{9.5}$$

where P_{in} is the average power input into the amplifier (which is about half of the peak power in the bit stream, assuming equal probability of 1's and 0's). For large gains G $\gg$ 1 and assuming that $B_o = 12.5$ GHz for a wavelength of 1550 nm, we obtain

$$\text{OSNR(dB)} \approx P_{\text{in}}(\text{dBm}) + 58 - \tilde{F} \tag{9.6}$$

where

$$\tilde{F}(dB) = 10 \log F = 10 \log 2n_{\text{sp}} \tag{9.7}$$

is referred to as the *noise figure* of the amplifier (for large gains). For $n_{\text{sp}} = 2$ and $P_{\text{in}} = -30$ dBm, we obtain an OSNR of 22 dB. This implies that the signal power is larger than the noise power by a factor of approximately 160 within the spectral width of 12.5 GHz. In system designs, typically one looks for an OSNR greater than 20 dB for the detection to have low bit error rates (BERs).

Each amplifier in a chain adds noise, and thus in a fiber optic communication system consisting of multiple spans of optical fiber links with amplifiers, OSNR will keep falling (Fig. 9.13) and at some point in the link when the OSNR falls below a certain value, the signal would need to be regenerated. If we assume a noise of 0.6 μW added by each amplifier, then after, say, 10 amplifiers, the signal power would still be the same as at the beginning (assuming the amplifier gain exactly

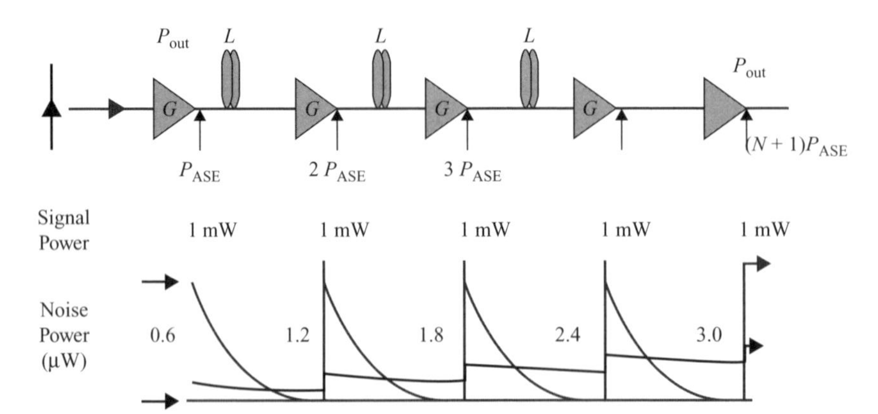

FIGURE 9.13 In a long-distance communication link with amplifiers spaced by spans of fiber length L, although the signal levels are maintained at the same level after every span, the noise power keeps increasing.

compensates for the attenuation in the span) but the noise power would be 6 μW. Thus, as the signal passes through multiple spans and amplifiers, there is a reduction in the optical signal-to-noise ratio. Hence there is a maximum number of amplifiers that can be placed in a link, beyond which the signal needs to be regenerated.

For a link consisting of multiple spans of transmission fiber and EDFAs compensating exactly for the loss of each span, the OSNR is given by

$$\text{OSNR(dB)} \approx \tilde{P}_{\text{out}}(\text{dBm}) - 10\,\log n + 58 - \tilde{F}(\text{dB}) - 10\,\log(N+1) - \tilde{L}_{\text{sp}}(\text{dB}) \tag{9.8}$$

where $\tilde{P}_{\text{out}}$ is the total output power from the amplifier in dBm, n represents the number of wavelength channels in the link, $\tilde{F}$ represents the noise figure of each EDFA (assumed to be the same), N represents the number of amplifiers, and $\tilde{L}_{\text{sp}}$ is the loss in decibels of each span.

As a typical example of the use of Eq. (9.8), let us consider a link consisting of EDFAs with the following specifications:

Amplifier output power	$\tilde{P}_{\text{out}} = 17\,\text{dBm}$
Number of channels	$n = 32$
Noise figure	$\tilde{F} = 5\,\text{dB}$
Span loss	$\tilde{L}_{\text{sp}} = 20\,\text{dB}$

If we require an OSNR of 22 dB at the end of the link, then, using Eq. (9.8), the maximum number of amplifiers that can be used in the link is about 18. If more than this number of amplifiers are employed, the OSNR will fall below the required value of 22 dB. Thus, for proceeding farther along the length, the signal needs to be regenerated.

It is also interesting to note from Eq. (9.8) that the OSNR increases by 1 dB if the output power $\tilde{P}_{\text{out}}$ from the amplifier increases by 1 dB or the noise figure $\tilde{F}$ decreases by 1 dB or the span loss $\tilde{L}_{\text{sp}}$ decreases by 1 dB. Thus, to achieve the same OSNR at the end of a multispan link, the number of permissible amplifiers in the chain can be increased by reducing the noise figure of each amplifier, increasing the output power of the amplifiers, or decreasing the span loss. Indeed, by choosing lower span loss, the number of amplifiers can be increased significantly so that the distance for regeneration can be made very large. Thus, reducing each span loss by 3 dB would result in a doubling of the maximum number of amplifiers allowed (all other parameters being the same). Of course, in this case, we would have to employ more amplifiers.

When an optical signal is detected by a photodetector, it generates a current that is proportional to the optical power falling on the detector. When an amplified signal falls on the photodetector, the light generated by the ASE falls simultaneously. The electric field of light waves corresponding to the signal and the ASE beat together to produce *beat noise*. Thus, although the average current generated by the detector would be proportional to the light power falling on it, the current will exhibit beat noise riding over the signal. This beating is very similar to the sound beats with which one is familiar. When two sound waves differing only slightly in frequency fall on our ears, we hear a waxing and waning of the sound at a frequency equal to the difference in frequency between the two sound waves. Similarly, in the case of light waves, the different frequencies present in the ASE can beat with the amplified signal to produce signal-spontaneous beat noise, while the various frequency components of the ASE can beat with other frequency components of the ASE to produce spontaneous–spontaneous beat noise. Apart from these, there are other noise components, called shot noise and thermal noise (discussed in Chapter 8). Thus, the signal exiting the EDFA is expected to be more noisy than the signal entering the EDFA. We define the *noise figure* (NF) of the amplifier as the ratio of the input electrical signal-to-noise ratio to the output electrical signal-to-noise ratio.

If the input signal is shot-noise limited (i.e., the noise is due primarily to shot noise), the electrical signal-to-noise ratio (SNR) of the input is given by

$$\mathrm{SNR}_{\mathrm{in}} = \frac{P_{\mathrm{in}}}{2h\nu B_e} \tag{9.9}$$

where B_e is the electrical bandwidth of the detector. Similarly, the SNR of the output signal after detection is given by

$$\mathrm{SNR}_{\mathrm{out}} = \frac{G P_{\mathrm{in}}}{2h\nu B_e 2n_{\mathrm{sp}}(G-1)+1} \tag{9.10}$$

Hence, the noise figure of the amplifier will be

$$F = \frac{\mathrm{SNR}_{\mathrm{in}}}{\mathrm{SNR}_{\mathrm{out}}} = \frac{1+2n_{\mathrm{sp}}(G-1)}{G} \tag{9.11}$$

For $G \gg 1$, $F \simeq 2n_{\mathrm{sp}}$.

An amplifier with $n_{\mathrm{sp}} = 2$ and $G = 100$ ($\tilde{G} = 20$ dB) would have a noise figure of $F = 4$ or in decibel units, $\tilde{F} = 6$ dB. This implies that when the signal gets amplified by the EDFA, the electrical signal-to-noise ratio will deteriorate by a factor of 4.

One may wonder why, when the EDFA is actually causing the signal-to-noise ratio to deteriorate amplifiers are used. If the input signal is very weak, it may not be possible to detect the signal if it falls below the sensitivity level of the detector, even if its signal-to-noise ratio was very high. The amplifier boosts the signal (and, of course, the noise, too) and brings it to a level that the detector can measure; of course, the price paid in this process is a deterioration of the signal-to-noise ratio since the amplifier also adds its own noise. As long as the signal-to-noise ratio is higher than a certain value, it will be possible to detect and process the signal. The noise figure is a very important characteristic of all amplifiers. Typical commercial amplifiers have a noise figure between 4 and 6 dB.

In the same context, we can define the noise figure of a lossy element such as a piece of fiber or splice. The output SNR in this case is given by

$$\mathrm{SNR}_{\mathrm{out}} = \frac{T P_{\mathrm{in}}}{2h\nu B_e} \tag{9.12}$$

where T (<1) is the power transmission coefficient (the ratio of the output power to the input power) of the element. Note that the lossy element does not add noise to the signal; it leads only to attenuation of the signal. Assuming the input SNR to be the same as given earlier (i.e, shot-noise limited), we get for the noise figure of a lossy element,

$$F = \frac{1}{T} \tag{9.13}$$

Thus, if an element has a transmission of 50% (i.e., $T = 0.5$, a loss of 3 dB; that is, the output power is half of the input power), the element can be said to have a noise figure of $F = 2$ or $\tilde{F} = 3$ dB. Hence, each lossy element would contribute to deterioration of the signal-to-noise ratio of the propagating signal.

When we have a chain of elements with different gains (or losses) and different noise figures, the noise figure of the chain is given by

$$F = F_1 + \frac{F_2 - 1}{G_1} + \frac{F_3 - 1}{G_1 G_2} + \frac{F_4 - 1}{G_1 G_2 G_3} + \cdots + \frac{F_N - 1}{G_1 G_2 G_3 \cdots G_{N-1}} \tag{9.14}$$

where G_i and F_i are the gain and noise figure of the ith amplifier. As is evident from Eq. (9.14), the overall noise figure of the chain is dominated by the noise figure F_1 of the first amplifier in the chain.

Equation (9.14) can be used to evaluate the overall noise figure of combinations of lossy elements and an amplifier. Thus, if we combine a lossy element with transmission T and an amplifier with gain G and noise figure F, if the lossy element follows the amplifier, the overall noise figure would be

$$F_1 = F + \frac{1/T - 1}{G} \tag{9.15}$$

If, on the other hand, the amplifier follows the lossy element, the overall noise figure would be

$$F_2 = \frac{1}{T} + \frac{F - 1}{T} = \frac{F}{T} \tag{9.16}$$

Although the overall gain in both cases would be same (neglecting any saturation effects), the overall noise figures are different in the two cases. As a typical example, if we consider an amplifier with a gain of 100 and a noise figure $F = 6$ (i.e., $\tilde{F} = 4$ dB) and a lossy element with $T = 0.5$, then $\tilde{F}_1 \approx 4$ dB and $\tilde{F}_2 \approx 7$ dB. This shows that any loss before amplification would cause the noise figure of the system to deteriorate. Hence, it is very important to reduce the losses at the entry point of an EDFA.

The analysis given above can be used for design of dual-stage amplifiers with lossy elements between the two stages. The lossy elements could be dispersion-compensating modules or add/drop multiplexers or any other element. Such amplifiers are referred to as *midstage access amplifiers*. It is in view of the discussion above that Raman amplifiers are characterized by equivalent negative noise figures; noise in Raman fiber amplifiers is discussed in Chapter 10.

Example 9.6 Let us consider a link with eight amplifiers with a span loss of 20 dB and an amplifier gain of 20 dB. Let us also assume that the ASE noise added by each amplifier is 0.6 μW within the signal band. At the end of each span the signal level is brought back to the input level of, say, 1 mW. On the other hand, let us look at the evolution of the noise. The first amplifier adds a noise of 0.6 μW, which when it propagates through the span and reaches the next amplifier will have a power of 6 nW. Since the amplifier gain is 20 dB, this noise will get amplified to 0.6 μW as it exits the amplifier. But note that the second amplifier will add its own noise of 0.6 μW, which implies that the noise exiting the second amplifier is now 1.2 μW. Thus, after each span the signal power is brought back to the initial value while the noise power keeps on building. After, say, 15 amplifiers, the noise power would be 9 μW. Thus, the optical signal-to-noise ratio at the end of 15 amplifiers would be 1 mW/9 μW = 111 ~ 20 dB, just sufficient for detection within specified bit error rates. This example clearly shows the buildup of noise with each span and the reduction in signal-to-noise ratio.

9.6 GAIN SATURATION

As for any amplifier, for small input signal powers, the gain of an EDFA remains almost constant. As the input signal power increases, the gain starts to decrease. This happens since as the signal power increases, the atoms in the excited state are used up faster, and with a fixed pumping rate, the pump is unable to replenish the upper state population, and this results in a reduction of the population inversion and hence the gain. This is referred to as *gain saturation*. Figure 9.14 shows a typical measured

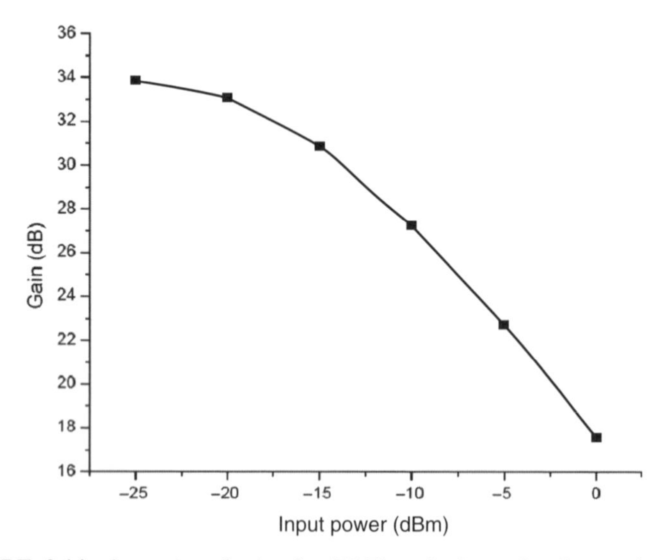

FIGURE 9.14 Saturation of gain of an EDFA as the input signal power increases.

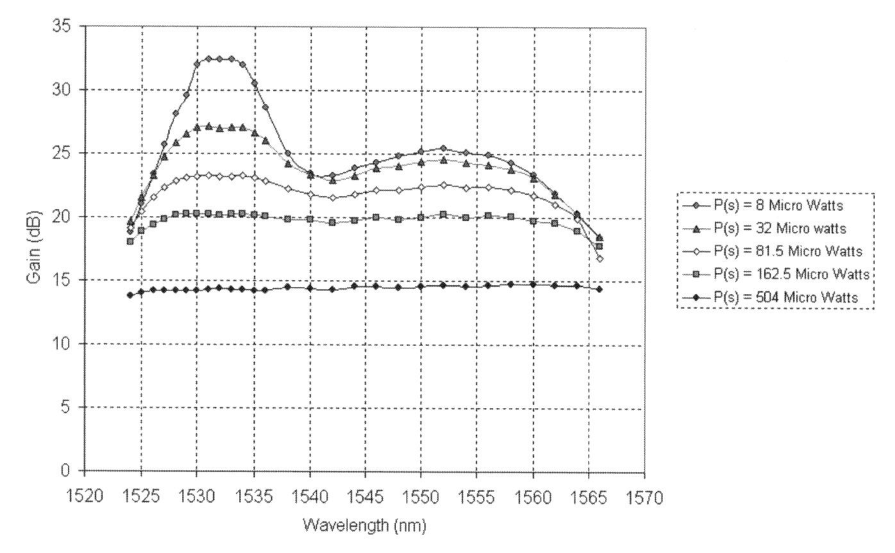

FIGURE 9.15 Spectral variation of gain for different input signal powers. Note that as the signal power increases, the gain is reduced and the spectral gain variation changes.

gain saturation effect in an EDFA. The gain saturation behavior of the amplifier can be approximated by the following equation:

$$G = G_0 \exp\left(-\frac{G-1}{G}\frac{P_{\text{out}}}{P_{\text{s}}}\right) \tag{9.17}$$

where G_0 is the unsaturated gain (gain for low output powers), P_{out} is the output power, and P_S is a constant called the *saturation power*. As Eq. (9.17) shows, as the output power increases, the gain reduces.

Figure 9.15 shows the measured spectral variation of gain for different input signal powers, clearly showing gain saturation with increase in the input signal power. Apart from saturation, the spectral gain profile also shows variations, and when an amplifier is to be used for varying input powers, this fact has to be taken into account in designing gain-flattening techniques.

A very important parameter of an EDFA is the *output saturation power*, which is defined as the output power when the gain of the amplifier is reduced by a factor of $\frac{1}{2}$ from the small-signal value (i.e., the output signal power for which the gain is $G_0/2$). From Eq. (9.17), the output saturation power can be evaluated as

$$P_{\text{out}}^{\text{sat}} = \frac{G_0 \ln 2}{G_0 - 2} P_S \approx 0.69 P_S \qquad (\text{for } \tilde{G} > 20\,\text{dB}) \tag{9.18}$$

For EDFAs the saturation output power is about 17 to 23 dBm (i.e., 50 to 200 mW): that is, when the output power from the EDFA is 50 mW (for a 17-dBm amplifier), the gain of the EDFA is reduced by a factor of 2 (= 3 dB) with respect to the gain when the signal power is low. If one tries to draw greater power by increasing the input signal power, the gain would decrease further.

Booster amplifiers operate in the saturated regime since the input powers are quite high, whereas preamplifiers operate at very low input powers and hence in the unsaturated region.

Example 9.7 Let us consider an EDFA with a small-signal gain of 23 dB and a saturation output power of 15 dBm. This implies that for low input signal powers, the amplifier gain would be 23 dB. For example, for an input of 1 μW, the output power would be 200 μW. A saturation output power of 15 dBm implies that when the output power is 15 dBm ($\simeq$ 31.6 mW), the gain of the amplifier is $23 - 3 = 20$ dB. This corresponds to an input power of 0.316 mW. Thus, as the input power level rises from 1 μW to 0.316 mW, the gain falls off from 23 dB to 20 dB. If the input power is increased further, the gain would decrease further.

9.7 GAIN TRANSIENTS

Since the signals propagating through the doped fiber in an EDFA deplete the energy stored in the population inversion to get amplified, the gain provided by the EDFA would depend on the input signal power. As the input signal power increases, the population inversion would decrease, resulting in reduced gain (gain saturation). When an EDFA is used in an optical network, then when channels are added (or dropped), the input power to an EDFA in the link would increase (decrease). This would result in a decrease (an increase) of gain of the surviving channels. Figure 9.16 shows the time variation of output power of an EDFA, in which of the eight input channels, each carrying 0.5 mW of power, seven channels are dropped at $t = 0$ and then added again a little later. When the channels are dropped, only one signal wavelength is input into the amplifier, resulting in increased amplifier gain. The gain increases with time and saturates after about 200 μs. When the channels are added, the gain reverts back to its old value, thus reducing the power in the surviving channel. Such gain transients can cause problems in a fiber optic communication link since increased output signal powers can saturate the detector and thus cause increased bit error rates. If the drop and add process is done rapidly, the gain excursion is quite low.

To maintain the surviving channels under existing conditions when channels are added or dropped, we need to keep their gain constant. This can be achieved by increasing (decreasing) the pump power as the gain decreases (increases) due to adding and droping channels. The time variation of gain in the surviving channels depends on various factors, and to keep the gain constant, one needs to provide electronic feedback to the pump so as to react to keep the gain constant. The feedback provided is fast enough that the gain transients are restricted within some maximum value. Typically, if half the input channels are dropped (i.e., the input signal power falls by 50%), the settling time is about 60 μs and the gain excursion is within 0.15 dB.

Note from Fig. 9.16 that if droping and adding channels is made faster and faster, the gain modulation decreases. This is due primarily to the large lifetime of the

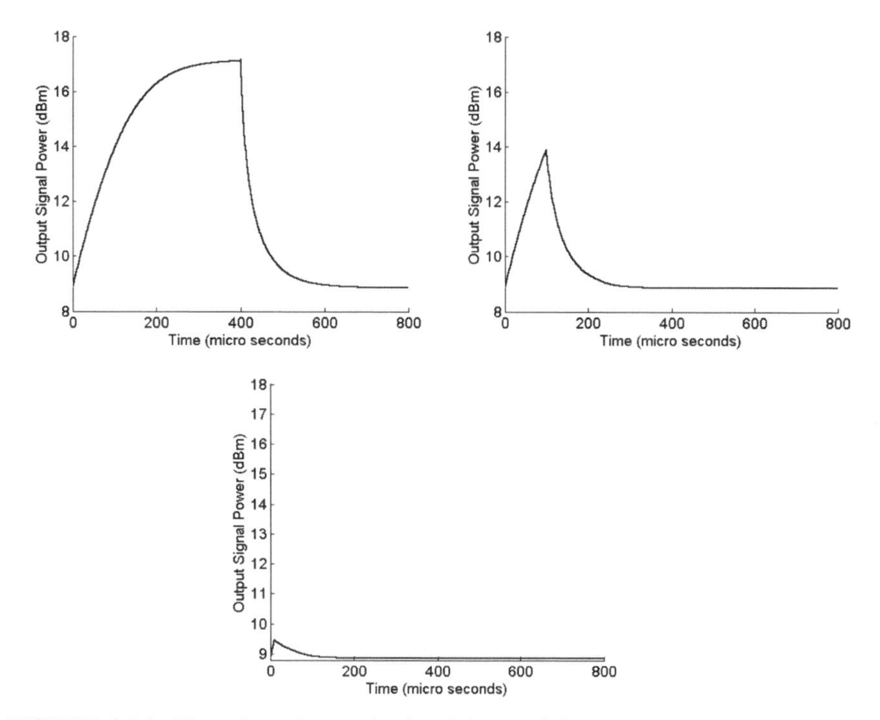

FIGURE 9.16 Time dependence of gain of the surviving channel when seven of eight channels are dropped at $t = 0$ and then added at different times.

excited level of erbium ions. In view of this, EDFAs do not suffer from cross-gain modulation (i.e., modulation of channel gain due to the presence of pulses in another channel).

9.8 EDFA MODULES

A number of vendors sell EDFAs with different operating characteristics. Table 9.1 gives typical specifications of a commercial EDFA. Figure 9.17 shows a typical EDFA module developed by IIT Delhi, Tejas Networks Pvt. Ltd, Bangalore and Optiwave Photonics, Hyderabad. The specifications of this model are given in Table 9.2.

9.9 FIBER LASERS

In an optical amplifier, an input signal in an appropriate wavelength range is amplified by the amplifier. If such an amplifier is provided with optical feedback (i.e., part of the output is reflected back into the amplifier), the device can act like a laser. In such a device the pump creates population inversion within the doped fiber, and some of the atoms taken into the excited state come down by emitting spontaneously. As

TABLE 9.1 Typical Specifications of Commercially Available EDFAs

Wavelength range (C-band)	1530 to 1565	nm
Wavelength range (L-band)	1570 to 1605	nm
Total output power	+23	dBm
Total input power range	−26 to +10	dBm
Gain	30	dB
Gain flatness	0.8	dB
Noise figure $P_{in} = 0$ dBm (C-band)	5.5	dB
Noise figure $P_{in} = 0$ dBm (L-band)	6.5	dB
Polarization-dependent gain	0.4	dB
Polarization mode dispersion	0.5	ps
Transient settling time	50	μs
Transient overshoot/undershoot	0.5	dB
Power consumption	20	W

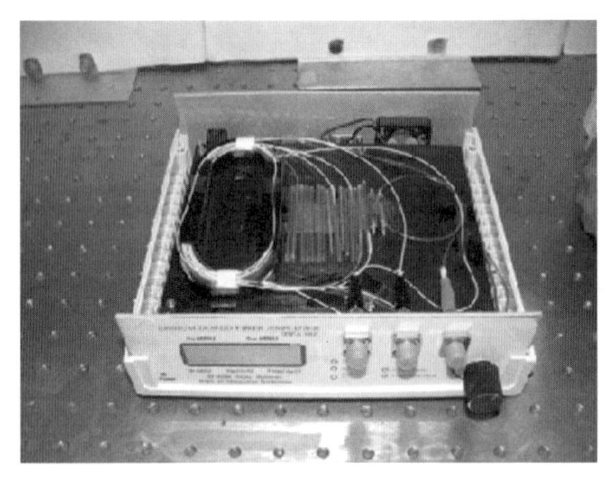

FIGURE 9.17 EDFA module developed by IIT Delhi, Tejas Networks Pvt. Ltd., Bangalore, and Optiwave Photonics, Hyderabad.

TABLE 9.2 Optical Specifications: IITD EDFA

Amplifier type	Booster
Operational wavelength	C-Band
Number of channels	One
Input signal power (P_{in})	−25 to −5 dBm
Total output power at $P_{in} = -25$ dBm	+10 dBm
Total output power at $P_{in} = -5$ dBm	+15 dBm
Small-signal gain	+34 dB
Noise figure	≤5 dB
Polarization-dependent loss	0.20 dB
Polarization mode dispersion	0.04 ps

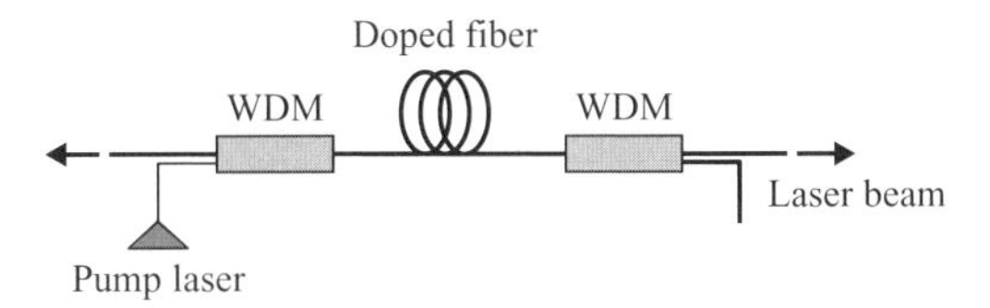

FIGURE 9.18 Typical fiber laser.

discussed earlier, some of the spontaneously emitted light couples into the fiber mode and propagates along the fiber. When it reaches the end, a part of it is reflected back into the population inverted fiber. This reflected light then undergoes amplification until it reaches the other end, and again, a part of the light is reflected back into the fiber. Light undergoing this to-and-fro oscillation within the cavity loses a part of the energy at every reflection and gains energy as it propagates through the doped fiber. If the energy lost by the light beam in one round trip is regained by the beam while propagating through the doped fiber in one round trip, this leads to laser oscillation. At this stage the pumped fiber acts as a source and becomes a fiber laser. Indeed, in 1961, Elias Snitzer wrapped a flashlamp around a glass fiber (having a 300-μm core doped with Nd^{3+} ions clad in a lower-index glass) and when suitable feedback was applied, the first fiber laser was born. Thus, the fiber laser was fabricated within an year of the demonstration of the first-ever laser by Theodore Maiman.

Figure 9.18 is a schematic of a fiber laser. As an example, let us consider an erbium-doped fiber laser. Light from a pump laser emitting at 980 nm is coupled into a short length of erbium-doped fiber using a wavelength-division-multiplexing (WDM) coupler. The WDM coupler at the output removes any unused pump laser power. If the two ends of the signal ports of the WDM coupler are cut properly and not treated with coatings, they can reflect about 4% of the incident light. Since the gains provided by erbium ions is very large, this small reflectivity is sufficient to satisfy the condition for laser oscillation: namely, compensation of loss by the gain provided by population inversion. This would result in an output laser beam from both ends of the coupler. The wavelength of emission is usually determined by the wavelength satisfying the maximum gain and minimum loss; this is around 1530 nm for erbium-doped fibers. Figure 9.19 shows the output from a erbium-doped fiber laser as the pumping is increased. Just before starting to lase, the pump power is insufficient to overcome the losses in the cavity, and thus the output is only amplified spontaneous emission (the lower curve in Fig. 9.19). As we increase the pump power, the erbium-doped fiber starts to lase and the spikes correspond to the various resonator modes; the ends of the fiber act as the resonator.

In case it is necessary to have the laser oscillate at a specific wavelength within the gain bandwidth of the erbium ion, then this can be achieved by using a fiber Bragg grating (FBG) at one end of the laser. Since an FBG reflects a particular wavelength, this wavelength would have much higher feedback into the laser cavity and thus would suffer much lower loss than would other wavelengths. This would ensure that the fiber laser oscillates at the frequency determined by the FBG. Fiber lasers possess many interesting advantages vis-à-vis other laser systems. In particular, since the laser beam is confined to a very small cross-sectional area within the core of the

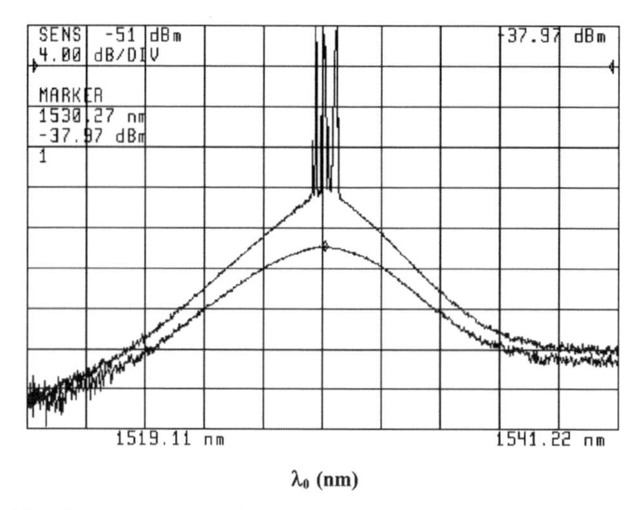

FIGURE 9.19 Output spectra of a fiber laser below and above threshold for laser oscillation.

fiber, large pump intensities can be achieved even with small pump powers, thus leading to lower pump power thresholds. Since both the pump and the laser beam are propagating within the fiber, they overlap very well, and this adds to increased laser efficiency and efficiencies of 80% are possible. Since the fiber guides the pump beam, one can use very long cavities without bothering about the divergence of the pump laser beam. Since the ratio of surface area to volume of fiber laser is very large, it does not suffer from thermal problems, and heat dissipation is much easier. The output beam is of very good quality since it emerges as the fundamental mode of the fiber. Also, since the components in the laser are made up of fibers that are all spliced, there are no mechanical perturbation problems as in bulk lasers with separate mirrors.

With conventional fibers with doped single-mode core and cladding, the laser power is restricted to about 1 W. To achieve higher output powers, fiber lasers use double-clad fibers as the amplifying medium (Fig. 9.20). In this fiber, the central core guides the laser wavelength and is single moded at this wavelength. The inner cladding is surrounded by an outer cladding, and this region acts as a multimoded

FIGURE 9.20 Double-clad fiber design for high-power fiber lasers.

guide for the pump wavelength. This radius of the inner cladding is large and so is the refractive index difference between the inner and outer cladding. This ensures that power from large-area diode lasers can be launched into the fiber efficiently. At the same time, since the laser wavelength is propagating as a fundamental mode in the inner core, the laser output would be single moded. The pump power propagating in the inner cladding propagates in the form of different rays (or modes). If the cladding is circular in cross section, it is possible that some of the rays propagating in the cladding never have an opportunity to cross the core, and this portion of the pump would never be used in creating inversion and thus leading to reduced conversion efficiencies. In order that all the rays corresponding to the pump power propagating in the inner cladding of the fiber cross the core, the inner cladding is made noncircular (Fig. 9.20). This type of design can lead to greatly increased pump conversion efficiencies.

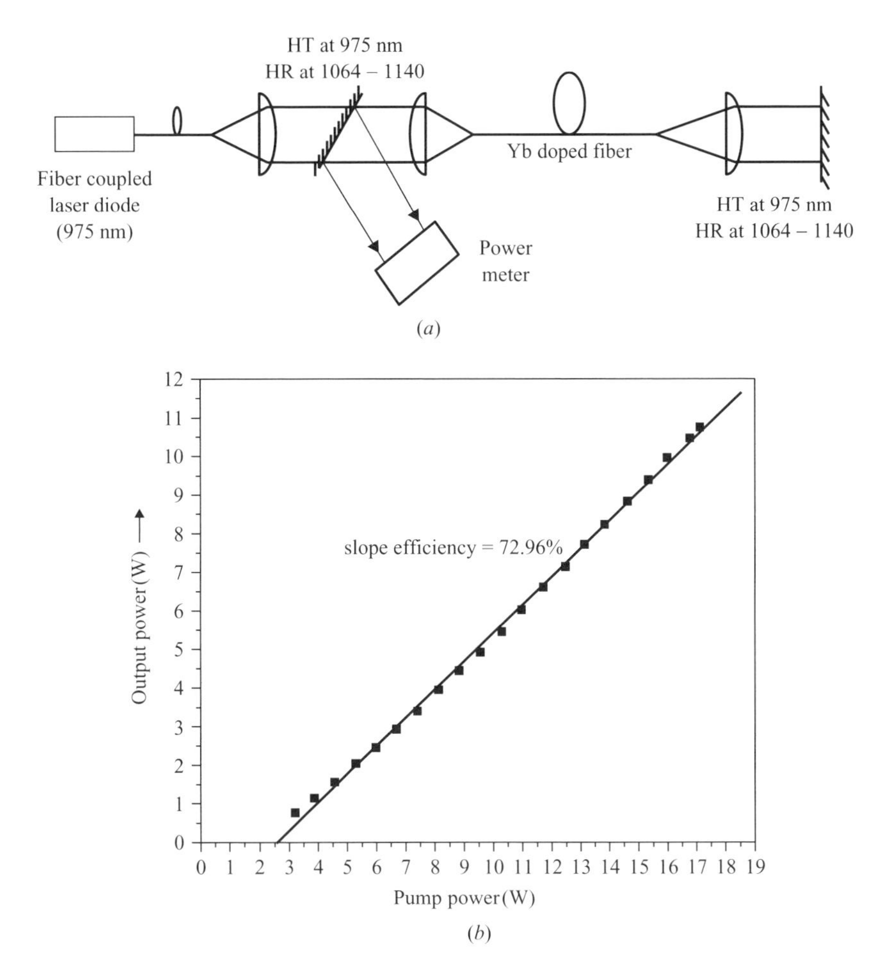

FIGURE 9.21 (*a*) High-power fiber laser; (*b*) laser power variation with input pump power. HT, high transmittivity; HR, high reflectivity. (After Upadhyaya et al., 2005.)

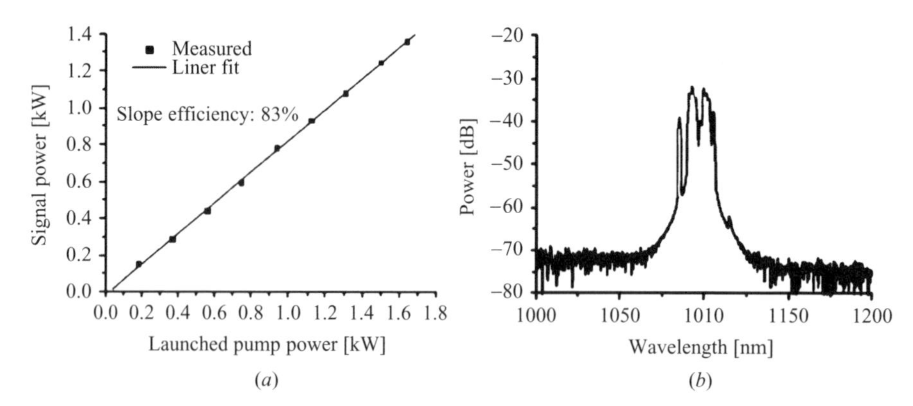

FIGURE 9.22 (*a*) Fiber laser output variation with input pump power, showing generation of 1.3 kW of continuous-wave laser output; (*b*) output laser spectrum. (Adapted from Jeong, 2004)

Figure 9.21*a* shows a fiber laser construction with pump power (at 975 nm) coupled through one end of the doped fiber and a dichroic mirror (which transmits the pump wavelength but reflects the laser wavelength), and Fig. 9.21*b* shows the corresponding laser output with input pump power. With a cladding pumped geometry continuous-wave fiber laser (ytterbium-doped fiber) emitting more than 1000 W of power has been realized. Figure 9.22 shows the variation of output power with launched pump power of a very high power continuous-wave fiber laser. Output powers of 1.36 kW have been achieved, and the output beam exiting the fiber has an excellent quality. The fiber is an ytterbium-doped double-clad silica fiber with the laser guided by the central core and the pumps propagating in a large D-shaped inner cladding. The output spectrum of the laser output is also shown in the figure.

Such high-power fiber lasers are expected to find wide applications in various industries, such as for cutting and hole drilling. Fiber lasers should make these applications quite convenient and practical. Using various mode-locking techniques, pulse widths as short as a few hundred femtoseconds have been realized. Such lasers find applications in various scientific and technological areas. One of the very important areas in which fiber lasers are having a very strong impact is in femtosecond ($1\ \text{fs} = 10^{-15}\ \text{s}$) pulse generation. To generate ultrashort pulses the standard technique of mode-locking is used. The mode-locked fiber laser operates typically at repetition rates of 80 to 110MHz. Since lasing is taking place within the fiber that guides the beam, the alignment of the laser is very stable, and the pulsed operation is self-starting. Typical mode-locked fiber lasers produce several milliwatts of power, which are then amplified to a few hundreds of milliwatts by a fiber amplifier. By using such techniques, pulse durations shorter than 100 fs are possible.

CHAPTER 10

Raman Fiber Amplifiers

10.1 INTRODUCTION

In Chapter 9 we discussed erbium-doped fiber amplifiers, which are the most important optical amplifiers. Since the EDFAs work on the principle of population inversion between energy levels of the erbium ion, the band of wavelengths that can be amplified by an EDFA is restricted. Current EDFAs operating in the C-band (1530 to 1565 nm) and the L-band (1565 to 1625 nm) are available commercially. Some EDFA designs operating in the S-band (1460 to 1530 nm) are also becoming available. At other bands, EDFAs cannot operate and one has to look for other amplifiers based on other dopants (instead of erbium), other materials (semiconductor optical amplifiers), or other effects. Raman fiber amplifiers (RFAs) are amplifiers falling in the last category. In these amplifiers one uses the phenomenon of Raman scattering to amplify optical signals. *The attractive features of RFAs are that they can be made to work in any wavelength band simply by choosing appropriate pump wavelengths, and they have a large bandwidth.* Apart from this, the link fiber can itself be used as the amplifier, and thus the signal gets amplified as it covers the distance along the communication link itself. Such amplifiers are also referred to as *distributed amplifiers*. RFAs have attracted considerable attention, and in this chapter we deal with the basic operating characteristics of RFAs.

10.2 RAMAN EFFECT

Raman scattering is perhaps easily understandable if we consider light to consist of photons (quanta of energy), with each photon having energy proportional to the frequency of the light wave. When a monochromatic light beam gets scattered by a transparent substance, one of the following may occur:

1. Over 99% of the scattered radiation has the same frequency as that of the incident light beam (Fig. 10.1); known as *Rayleigh scattering*, this was discussed briefly in Chapter 5. The sky looks blue because of Rayleigh scattering, and

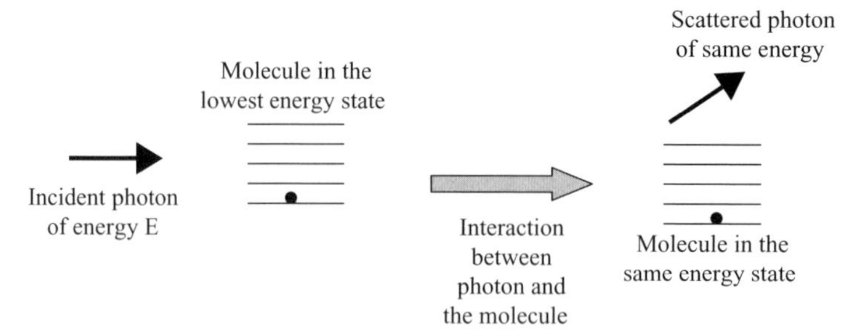

FIGURE 10.1 In Rayleigh scattering the energy of the incident photon remains unchanged as it interacts with the molecule and gets scattered in a random direction.

the light that comes out from the side of the optical fiber (Fig. 4.4) is also due to Rayleigh scattering.

2. A very small portion of the scattered radiation has a frequency that differs from that of the incident beam. This may arise due to one of the following three processes:
   (a) A part of the energy of the incident photon may be absorbed to generate translatory motion of the molecules. This would result in scattered light having a very small shift of frequency, which is usually difficult to measure. This scattering process is known as *Brillouin scattering*.[1]
   (b) A part of the energy E of the incident photon is taken over by the molecule in the form of rotational (or vibrational) energy, and the scattered photon has a smaller energy, E'. This leads to the scattered light having a less energy and hence lower frequency. The resulting spectral lines are known as *Raman Stokes lines* (Figs. 10.2*a* and 10.3).
   (c) On the other hand, the photon can undergo scattering by a molecule that is already in an excited vibrational or rotational state. The molecule can deexcite to one of the lower-energy states, and in the process, the incident photon can take up this excess energy and come out with a higher frequency. This leads to the scattered light having more energy and hence higher frequency. The resulting spectral lines are known as *Raman anti-Stokes lines* (Figs. 10.2*b* and 10.3).

The difference energy, which is $(E - E')$ for the Raman Stokes line and $(E' - E)$ for the Raman anti-Stokes line, would therefore correspond to the energy difference between the rotational (or vibrational) energy levels of the molecule and would hence be a characteristic of the molecule itself. The quantity $(E - E')$ or $(E' - E)$, usually referred to as the *Raman-shift* (Fig. 10.3), is independent of the frequency of the incident radiation. Through a careful analysis of the Raman spectra, one can

[1]The shift in frequency is usually measured in wavenumber units, which is defined later in this section. In Brillouin scattering the shift is $\lesssim 0.1\ \text{cm}^{-1}$, whereas in Raman scattering the shift is $\lesssim 10^4\ \text{cm}^{-1}$.

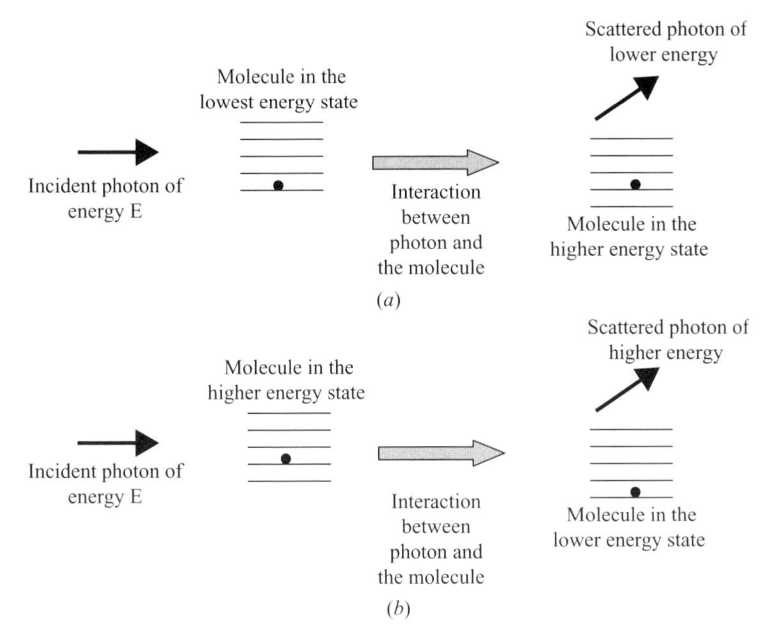

FIGURE 10.2 In the case of Raman scattering, during the interaction between the incident photon and the molecule, an exchange of energy takes place. (*a*) In Stokes scattering, a part of the energy of the photon is used to excite the molecule into an excited state, resulting in the output of a photon of lower energy. (*b*) In anti-Stokes scattering, energy from the molecule is added to the incident photon, resulting in a photon of higher energy.

determine the structure of molecules; there lies the tremendous importance of the Raman effect.

Since the probability of finding molecules is higher in the lower-energy state than in the higher-energy state, Stokes emission (into a lower-energy photon) is much more intense than anti-Stokes emission (into higher energy). Figure 10.3 shows a typical spectrum of scattered light observed showing Rayleigh scattering at the incident frequency, Stokes lines at lower frequency, and much weaker anti-Stokes lines at a higher frequency. Note that the frequency shifts in Stokes and anti-Stokes are equal; both involve the same set of energy levels of the molecule; however, anti-Stokes lines are much weaker than Stokes lines.

Figure 10.4*a* shows the Raman-scattered spectrum from a mixture of hydrogen and deuterium when the mixture is illuminated by a laser beam at 488-nm wavelength, the corresponding frequency being approximately 615 THz ($= 6.15 \times 10^{14}$ Hz). Now there is an energy level corresponding to vibration of the hydrogen molecule, which is separated from the lowest level by 125 THz ($= 1.25 \times 10^{14}$ Hz); thus, the Stokes line should appear at a (lower) frequency of 490 THz (690-nm wavelength) and the very weak anti-Stokes line should appear at a (higher) frequency of 740 THz (405 nm). In spectroscopy the frequency shift is measured in wavenumber units (i.e., inverse

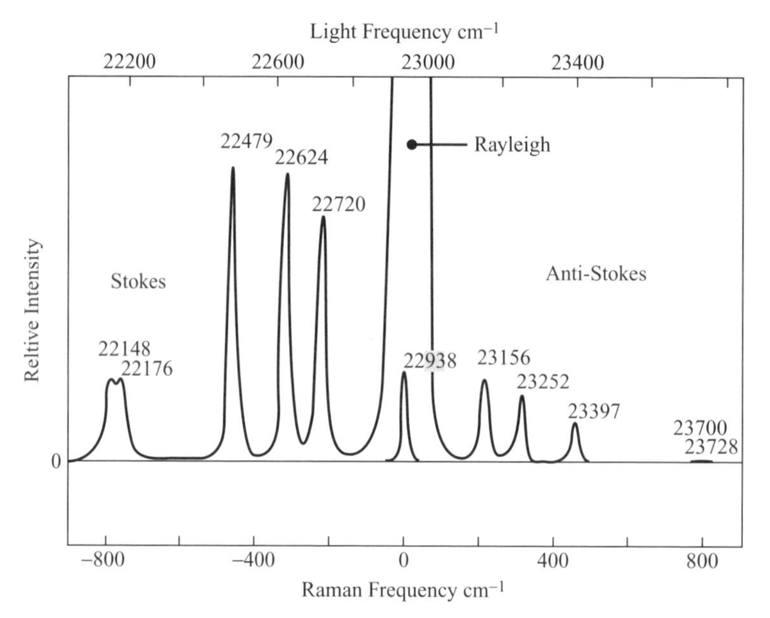

FIGURE 10.3 Scattered spectrum showing Rayleigh scattering (at the same frequency as the incident wave), Stokes emission (at a smaller frequency), and anti-Stokes emission (at a higher frequency). The Raman-scattered lines are characteristic of the energy levels of the molecule and are independent of the incident frequency. (Adapted from http://neon.otago.ac.nz/chemlect/chem306/pca/IR_Raman/page8.html.)

wavelength) and the units are usually cm^{-1} (read as "inverse centimeters"):

$$\Delta T(\text{cm}^{-1}) = \frac{\Delta E}{hc}$$

where ΔE is measured in ergs (1 erg $= 10^{-7}$ J), $h \approx 6.634 \times 10^{-27}$ erg·sec is Planck's constant, and $c \approx 3 \times 10^{10}$ cm/s is the speed of light in free space. Thus, the frequency shift in hertz will be given by

$$\Delta\nu(\text{Hz}) = \frac{\Delta E}{h} = \Delta T(\text{cm}^{-1})(3 \times 10^{10})$$

Thus, a frequency shift of 4155 cm^{-1} in wavenumbers corresponds to 125 THz, which is consistent with the value given above (Fig. 10.4*a*). In Fig. 10.4*a* the Raman shift due to D_2 and HD molecules is also shown—once again, the frequency shift is a "signature" of the molecule and is determined by the energy difference of the vibrational levels of the molecule. Since deuterium is heavier than hydrogen, the vibrational frequency of HD is smaller than that of hydrogen, and that of D_2 is even smaller. This is shown clearly in Fig. 10.4*a*, where the frequency shift of HD is

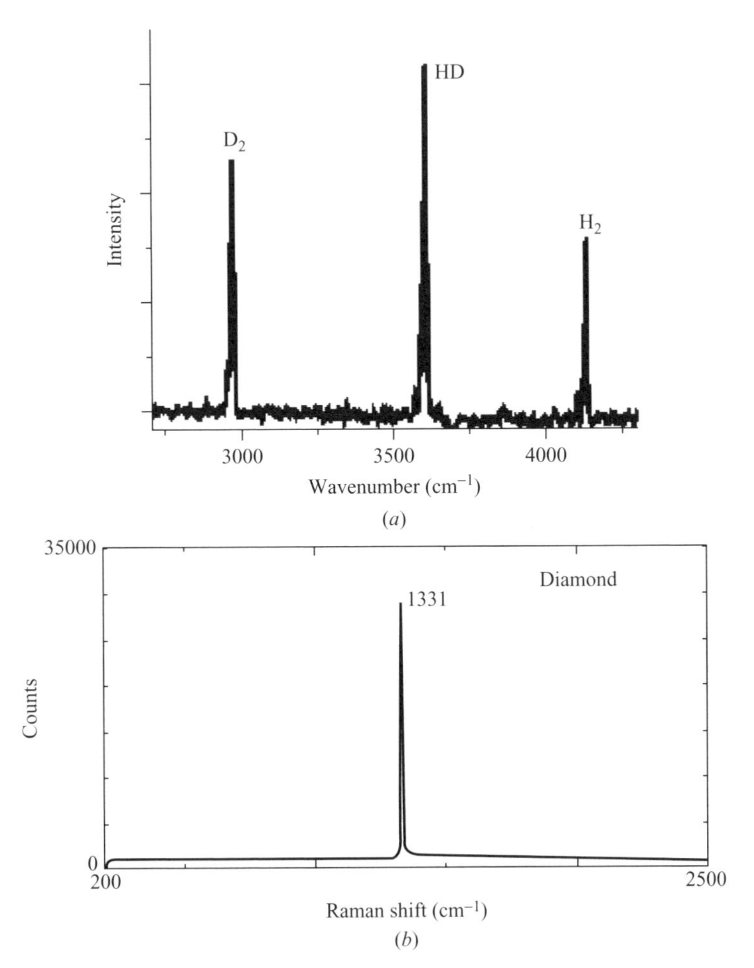

FIGURE 10.4 (*a*) Raman spectrum from a mixture of hydrogen and deuterium. The horizontal axis is the difference in frequency between the incident radiation and the scattered radiation. The hydrogen peak appears at a frequency shift of 4155 cm^{-1}, which corresponds to a frequency shift of 1.25×10^{14} Hz. The incident light wavelength was 488 nm, and hence the hydrogen peak will appear at a wavelength of 612.2 nm. (After Zeigler et al., 2003.) (*b*) Raman spectrum of diamond.

smaller than that of H_2, and that of D_2 is even smaller. Similarly, Fig. 10.4*b* shows the Raman-scattered spectrum of diamond, showing a Raman shift of 1331 cm^{-1}, which for a pump wavelength of 514 nm would correspond to a wavelength of about 549 nm.

Raman observed the shift of frequency due to scattering using the 4046-Å lines of a mercury lamp; Fig. 10.5 shows the actual Raman spectrum of the CCl_4 molecule as observed by Raman. The photograph is adapted from the 1930 Nobel lecture of C.V. Raman. It may be of interest to mention that on February 28, 1928, Raman and K. S.

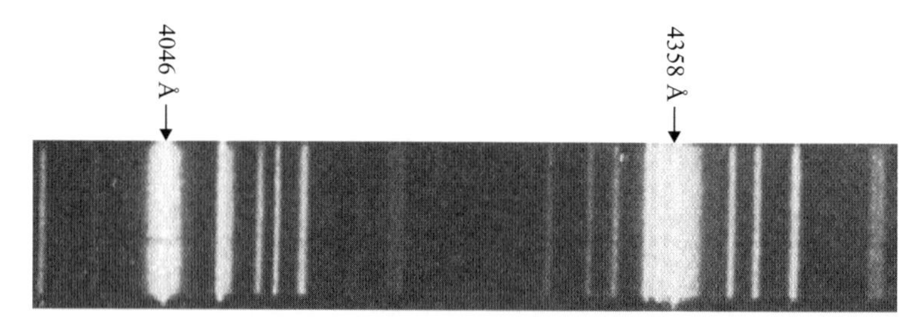

FIGURE 10.5 Raman spectra of CCl_4 observed for the 4046- and 4358-Å lines of a mercury lamp. (Adapted from the 1930 Nobel lecture of C. V. Raman.)

Krishnan observed the "Raman effect" in several organic vapors (e.g., pentane), which they called the "new scattered radiation." Raman made newspaper announcements on February 29 and on March 8, 1928, he communicated a paper entitled "A Change of Wavelength in Light Scattering" to *Nature*; the paper was published on April 21,1928. Although in the paper he acknowledged that the observations were made by Krishnan and him, the paper had Raman as the author and therefore the phenomenon came to be known as Raman effect, although many scientists (particularly in India) refer to it as the Raman–Krishnan effect. Subsequently, several papers were written by Raman and Krishnan. Raman got the Nobel prize in 1930 for "his work on the scattering of light and for the discovery of the effect named after him." At about the same time, Landsberg and Mandel'shtam (in Russia) were also working on light scattering, and according to Mandel'shtam, they observed "Raman lines" on February 21, 1928. But the results were presented in April 1928 and it was not until May 6, 1928 that Landsberg and Mandel'shtam communicated their results to the journal *Naturwissenschaften*—but by then it was too late! Much later, scientists from Russia were still calling Raman scattering "Mandel'shtam–Raman scattering." For a very nice historical account of the Raman effect, we refer the reader to a book by Venkataraman (1994).

In 1958, thirty years after the discovery of the Raman effect, Raman wrote an article on the Raman effect for *Encyclopaedia Britannica*. In that article he wrote: "The rotations of the molecules in gases give more readily observable effects, viz., a set of closely spaced but nevertheless discrete Raman lines located on either side of the incident line. In liquids, only a continuous wing or band is usually observed in the same region, indicating that the rotations in a dense fluid are hindered by molecular collisions. The internal vibrations of the molecules, on the other hand, give rise in all cases to large shifts of wave length. The Raman lines attributed to them appear well separated from the parent line and are therefore easily identified and measured."

The Raman spectra shown in Figs 10.3 and 10.4 are very sharp: that is, the Raman scattered light appears at well-defined frequencies due to the definite energy levels of the molecules with which the incident light beam interacts. In the case of an optical fiber, since it is made of fused silica (silicon dioxide) and the medium is amorphous (not crystalline), the Raman-scattered spectrum is broad.

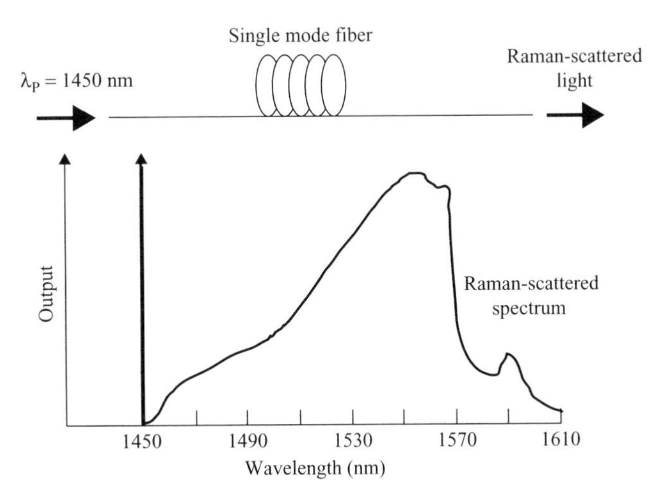

FIGURE 10.6 When light at 1450 nm is sent through a long optical fiber, in addition to the attenuated light at 1450 nm, we also observe light emanating at longer wavelengths. This is due to Raman scattering.

Now let us consider an experiment in which we send a strong light beam at a wavelength of 1450 nm through a long ($\simeq$ 10 km) optical fiber. The light beam undergoes Raman scattering from the molecules of the glass fiber, and this is expected to give rise to light appearing at higher wavelengths. Figure 10.6 shows a typical spontaneous Raman spectrum from an optical fiber pumped by radiation at 1450 nm. As can be seen, the scattered radiation occupies a large band, and the peak of the scattered radiation lies about 100 nm away from the pump wavelength. Indeed, Raman scattering in silica leads to a Raman shift of between 13 and 14 THz, which corresponds to about 100 nm at a wavelength of 1550 nm. As can be seen, the Raman spectrum from the optical fiber is not sharp but is broadened.

10.3 PRINCIPLES OF THE RAMAN FIBER AMPLIFIER

When we launch a high-power light beam into an optical fiber, we observe the appearance of Raman-scattered light at the end of the fiber referred to as *spontaneous Raman scattering*. If in addition to the strong pump light we launch a weak light beam (referred to as a *signal beam*), with its wavelength lying within the band of spontaneous Raman scattering, it leads to what is referred to as *stimulated Raman scattering* (SRS). In this case, the pump and signal wavelengths are coupled coherently by the Raman scattering process and the scattered radiation, is coherent with the incident signal radiation, much like stimulated emission that occurs in the case of a laser. The coherent nature of the process implies that the incident light gets coherently amplified by SRS. It is this process that is used to build Raman fiber amplifiers. Since the spontaneous Raman scattering spectrum is broad, the corresponding gain spectrum

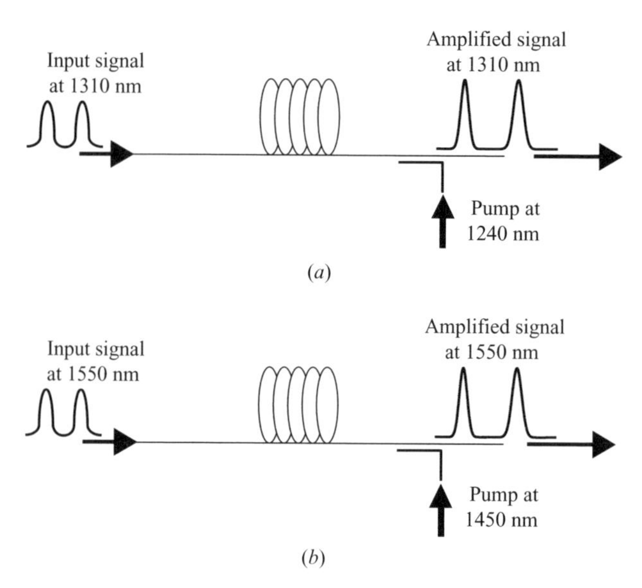

FIGURE 10.7 Raman amplifiers usually operate with backward pumping, wherein the signal and pump propagate in opposite directions. Using backward pumping the noise present in the pump does not get transferred to the signal, and this is the preferred pumping configuration. (*a*) Using a 1240-nm pump, wavelength signals at 1310 nm can be amplified. (*b*) Using the same fiber if the pump wavelength is changed to 1450 nm, 1550-nm wavelength signals can be amplified.

of the Raman amplifier is also very broad. The other interesting feature is that no matter what the wavelength of the pump light is, the fiber can act like an amplifier in the wavelength range corresponding to the spontaneous Raman scattering spectrum. Hence, if we need to amplify signals in the 1310-nm (which corresponds to 229 THz) window, we need to choose a pump wavelength of about 1240 nm (which corresponds to 242 THz), which will give a peak Raman scattering at a wavelength of 1310 nm, and such a pump will lead to amplification of signals at 1310 nm. Similarly, if we need to amplify signals in the 1550-nm (which corresponds to 194 THz) window, we need to choose a pump wavelength of about 1450-nm (which corresponds to 207 THz); in each case the pump frequency is about 13 THz more than the signal frequency (Fig. 10.7). Notice that unlike EDFA, which operated only in specific wavelength bands, Raman amplifier can operate in any wavelength region.

In Raman amplifiers the pump beam can propagate in the same direction as the signal, or in the reverse direction. The former case is referred to as *co-propagating* (forward pumping) and the latter as *contra-propagating* (backward pumping). The Raman scattering phenomenon is an extremely fast process with time scales in the femtosecond (10^{-15} s) regime. This can lead to transfer of power fluctuations from the pump to the signal. One way to avoid this is to have backward pumping (Fig. 10.7), wherein the pump fluctuation–induced gain fluctuations get averaged out, and thus the noise in the signal due to pump fluctuations is much lower.

In a Raman fiber amplifier one defines an on–off gain G, which is the ratio of output signal power in the presence of Raman pump to that of the output signal power in the absence of the Raman pump. An approximate relation for the on–off gain of a Raman fiber amplifier is

$$G(\text{dB}) = 10 \log \frac{P_s(L)}{P_s(0)e^{-\alpha_s L}} \approx 4.34 \frac{\gamma_R}{\alpha_p} P_p(L) \tag{10.1}$$

where $P_s(0)$ is the input signal power, $P_s(L)$ is the output signal power, α_p and α_s are the background attenuation coefficients at the pump and signal wavelengths, respectively, and γ_R is the Raman gain efficiency, defined by

$$\gamma_R = \frac{g_R}{A_{\text{eff}}} \tag{10.2}$$

which is usually expressed in units of $W^{-1}\ km^{-1}$. Here A_{eff} is the effective mode area of the fiber and g_R is the Raman gain coefficient, which is specific for a given material composition of the fiber. Figure 10.8 shows the spectral variation of g_R for pure silica for which the peak value is about 10^{-13} m/W. The gain bandwidth is approximately 40 THz wide, and the peak value of the gain coefficient is inversely proportional to the pump wavelength. In Eq. (10.1) the last approximation is valid for $\alpha_p L \gg 1$. Equation (10.1) neglects depletion of the pump due to Raman scattering, although depletion due to attenuation is taken into account. This approximation is reasonably good if the input signal powers are not very large.

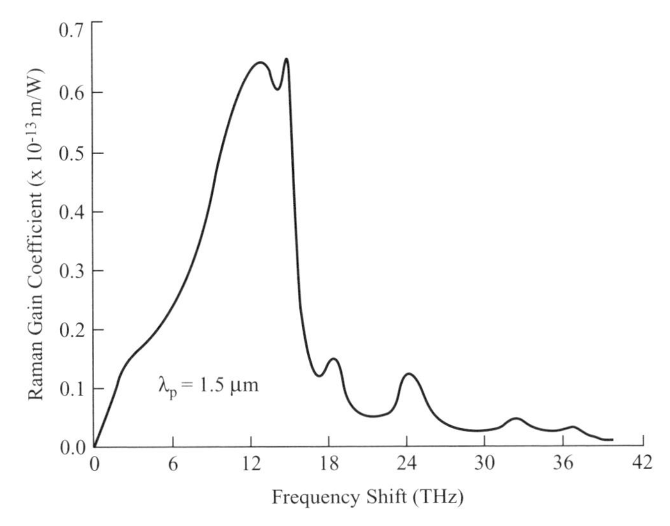

FIGURE 10.8 Spectral dependence of Raman gain coeffient g_R for a fused silica fiber. (Adapted from Islam, 2002. Copyright © 2002 IEEE.)

TABLE 10.1 Typical Values of γ_R for Various Fiber Types

Fiber Type	γ_R (W^{-1} km^{-1})
Standard SMF	0.5–1
Dispersion-compensating fiber	2.5–3
Highly nonlinear fiber	7.2
Photonic crystal fiber	8.8

If we launch light into an optical fiber, the intensity of light within the core of the fiber would depend on the transverse area occupied by the propagating light beam. Thus, fibers with smaller mode areas would provide higher intensity levels for the same light power. Since higher intensity would result in a higher magnitude of Raman scattering, the Raman gain efficiency would depend inversely on the effective area of the fiber mode [Eq. (10.2)]. Thus, to achieve large Raman gain, the effective area of the fiber should be small.

Typical values of γ_R for different types of fibers are given in Table 10.1. Photonic crystal fibers and holey fibers can have extremely small mode effective areas, and hence can provide extremely large Raman gains, as apparent from Table 10.1. Also, dispersion-compensating fibers (see Chapter 7) have smaller mode areas and hence possess larger Raman gain coefficients. Using DCFs with a Raman pump could indeed lead to zero-loss DCFs.

Let us consider a fiber with a Raman gain efficiency of 0.7 W^{-1} km^{-1} and a loss coefficient of 0.25 dB/km at the pump wavelength. In units of km^{-1}, a loss coefficient of 0.25 dB/km corresponds to $\alpha_p = 5.76 \times 10^{-2}$ km^{-1} (see Sec. 5.2). Since $1/\alpha_p \approx$ 17 km, if the fiber length is much larger than 17 km, the effective length of the fiber is 17 km (effective length is defined in Eq. 13.6). Assuming an actual fiber length of 80 km and a Raman pump power of 400 mW, the corresponding on–off gain would be [from Eq. (10.1)] about 21 dB. If the fiber attenuation at the signal frequency is also 0.25 dB/km, then for an input signal power of 0 dBm (= 1 mW) the output signal power in the absence of the Raman pump would have been -20 dBm, which is 10 μW. With a Raman pump power of 400 mW, the output signal power would be $+1$ dBm, which corresponds to a power level of 1.25 mW.

Figure 10.9 shows the simulated Raman gain spectrum of 100-km-long single-mode fiber pumped in the backward direction with 1 W at 1450 nm. Figure 10.10 shows the variation of the signal power at 1550 nm and the pump power at 1450 nm as a function of distance along a standard single-mode fiber in the backward-pumping configuration. The pump power is assumed to be 400 mW. In the absence of the pump, the signal gets attenuated to 10 μW at the end of the link. On the other hand, in the presence of the pump, the signal initially gets attenuated, but as it propagates toward the end of the link, it encounters the high-power pump and gets amplified via stimulated Raman scattering and exits with a power of 12 mW, leading to a net gain of 10.8 dB instead of a loss of 20 dB.

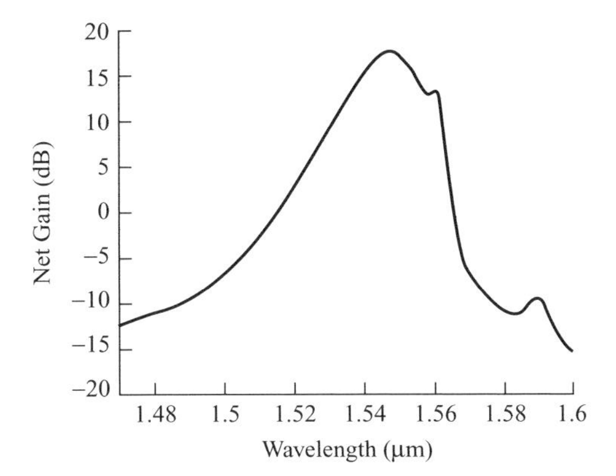

FIGURE 10.9 Simulated Raman gain spectrum of 100-km-long single-mode fiber pumped in a backward direction with 1 W at 1450 nm.

Figure 10.11 shows a measured on–off gain spectrum of a backward-pumped Raman fiber amplifier corresponding to an input pump power of 750 mW at a wavelength of 1453 nm and an input signal power of 0.14 mW over a 25-km-long span of single-mode fiber. On–off gains of greater than 12 dB are easily achievable. Since the gain spectrum depends on the pump wavelength, it is indeed possible to achieve a large flat gain spectrum using multiple pumps. Thus, using 12 pumps with wavelengths lying between 1410 and 1510 nm, a total 100-nm flat gain bandwidth from 1520 to 1620 nm (covering the C- and the L-bands) has been demonstrated (Fig. 10.12).

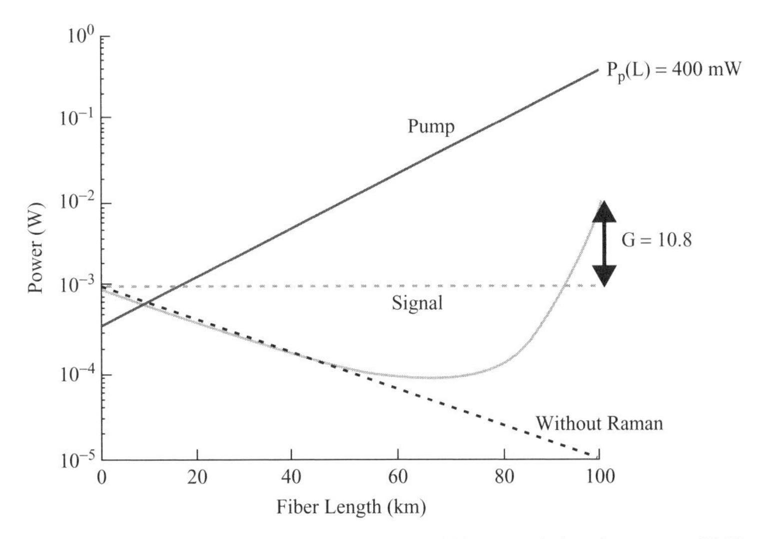

FIGURE 10.10 Variation of the pump power at 1450 nm and signal power at 1550 nm as a function of distance along a standard single-mode fiber in the presence of a pump. The pump is traveling in a backward direction.

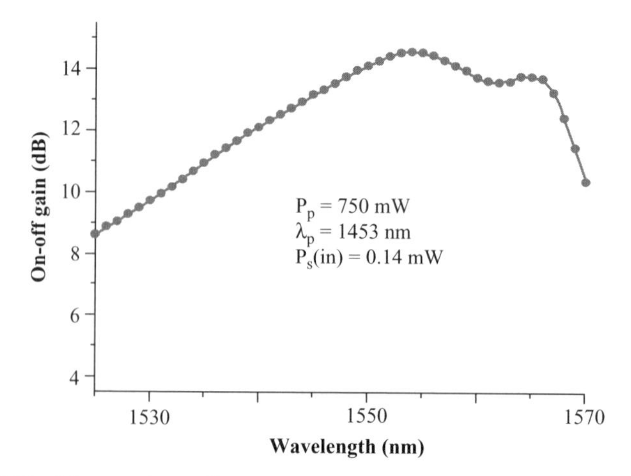

FIGURE 10.11 Measured on–off Raman gain spectrum of a 25-km-long single-mode fiber pumped by 750 mW of pump power at 1453 nm.

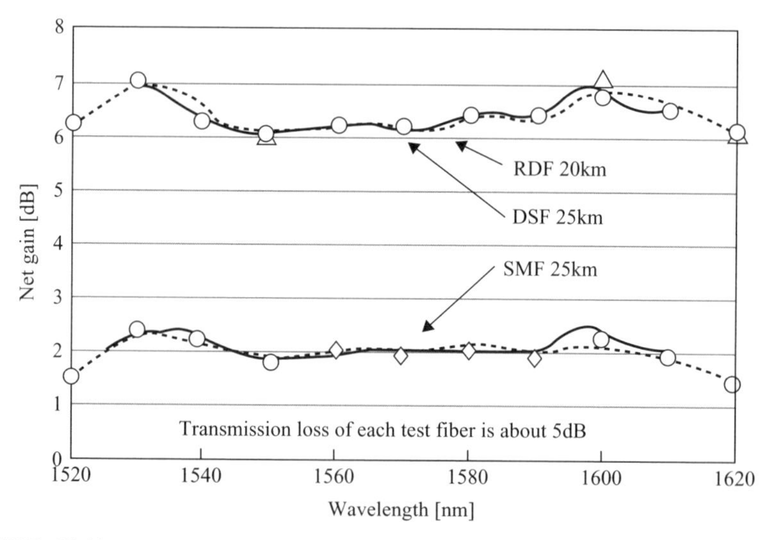

FIGURE 10.12 Raman gain spectrum using 12 pumps at different wavelengths and powers chosen so as to have almost flat gain over a large band of wavelengths. (Adapted from Emori and Namiki, 1999.)

10.4 NOISE IN RAMAN AMPLIFIERS

Due to the presence of a high-power pump, spontaneous Raman scattering takes place along with amplification of the signal. Since the spontaneous scattering has no relationship with the signal, it leads to noise in the amplifier. An approximate expression for the power emitted spontaneously by a Raman amplifier is given by

$$P_{\mathrm{sp}}(L) = 2h\nu\Delta\nu\left[\frac{4.34}{G}\exp\left(\frac{G}{4.34}\right) - \left(1 + \frac{4.34}{G}\right)\right] \tag{10.3}$$

where $\Delta\nu$ is the optical bandwidth. This radiation emitted spontaneously at the signal frequency constitutes the noise of the amplifier.

As an example, if we consider an on–off gain of 20 dB ($G = 100$), and if the optical bandwidth corresponds to 0.1 nm (corresponding to a bandwidth in the frequency domain of 12.5 GHz at 1550 nm), the noise power would be about 10 nW. This power gets generated all along the fiber as the pump propagates.

For an EDFA, the signal is amplified after it has traversed the link fiber. By contrast, in a distributed Raman amplifier, pump light is sent backward from the receiver end of the link, and thus as shown in Fig. 10.7, as the signal propagates it undergoes amplification before reaching the end of the link. In Section 9.5 we obtained the noise figure of a lossy element combined with an amplifier. As we discussed, any loss before amplification results in a deterioration of the noise figure. In the case of Raman amplifiers, signal amplification begins even before the signal has accumulated all the loss and thus are expected to have lower noise figures than EDFAs.

A distributed Raman amplifier can be replaced by an equivalent EDFA by assuming that the amplification takes place after propagation through the link. Referring to Fig. 10.13*a*, we represent the Raman amplifier by a lumped amplifier having a gain of G and a noise figure F with the lumped amplifier being followed by a fiber with a transmission coefficient T. The corresponding equivalent EDFA is assumed to have the same gain G and an equivalent noise figure F_{eq}. Using Eq. (9.13), we have for the case of a Raman amplifier (Fig. 10.13*a*)

$$\mathrm{NF} = F + \frac{1/T - 1}{G} \tag{10.4}$$

The corresponding noise figure of the equivalent EDFA is (see Eq. 9.16)

$$\mathrm{NF} = \frac{F_{\mathrm{eq}}}{T} \tag{10.5}$$

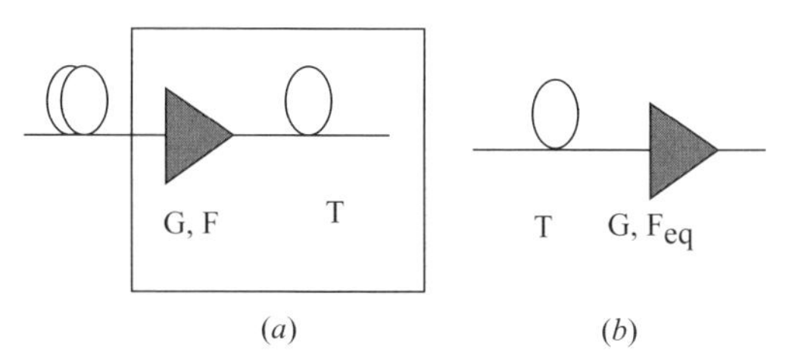

FIGURE 10.13 Equivalent EDFA for a Raman amplifier. The equivalent EDFA has a negative noise figure.

Comparing Eqs. (10.4) and (10.5), we obtain

$$F_{\text{eq}} = T\left(F + \frac{1/T - 1}{\text{G}}\right) \tag{10.6}$$

As an example, if we consider $G = 100, F = 2\,(\tilde{F} = 3\text{ dB}), T = 0.1$, the equivalent noise figure is $F_{\text{eq}} = 0.21$ ($\tilde{F}_{\text{eq}} = -6.8$ dB). Thus, we can represent a Raman amplifier by an equivalent noise figure which is negative. This does not, of course, imply that there is an improvement in the signal-to-noise ratio after amplification. The equivalent noise figure is that of an equivalent EDFA.

10.5 APPLICATIONS OF RAMAN FIBER AMPLIFIERS

As discussed earlier, since Raman fiber amplifiers can operate at any signal wavelength region, they allow us to expand the operation region of fiber optic communication systems to bands in which EDFAs do not operate. Apart from this, Raman amplifiers can also be used to extend the operation of optical fiber communication systems in the C-band. As an example, consider a multichannel fiber optic system consisting of 16 signals, each with a power of 2 mW (= 3 dBm). For a typical receiver sensitivity of 1.6 μW (= – 28 dBm), assuming a fiber attenuation of 0.25 dB/km, the signals can travel a distance of approximately 124 km (Fig. 10.14*a*). In some situations it may be necessary to go farther (e.g., due to the terrain in which the fiber system is being laid or because it requires crossing from one island to another). In such a case, it may be required to go a longer distance without putting any amplifier in between. Raman fiber amplifiers can indeed help in extending the link length. Thus, if a pump is sent in the backward direction from the receiver end of the fiber system, then by choosing the pump power appropriately, we can provide an additional (Raman) gain of, say, 15 dB, which would then permit us to have a hop length of 180 km rather than 124 km (Fig. 10.14*b*). Similarly, when the bit rate of a communication channel is increased, then for the system to operate without signal degradation (within a certain bit error rate), the receiver would need more power. In this case, the additional Raman gain that is realized by propagating an appropriate pump with the signal can be used to increase the transmission bit rate for the same distance between the transmitter and the receiver.

Since the gain coefficient depends on the effective area of the fiber, the Raman gain spectrum could be modified by proper fiber designs with appropriate spectral dependence of effective area and leakage loss. Thus, recently, novel fiber designs have been proposed that can give flat Raman gain with just a single pump [see Thyagarajan and Kakkar (2004) and Kakkar and Thyagarajan (2005)]. Figure 10.15 shows the gain spectrum of the novel segmented clad fiber with a single Raman pump at 1450 nm pumped in both the forward and backward directions. The fiber has also been shown to possess high dispersion value and thus can compensate for dispersion accumulated in the fiber.

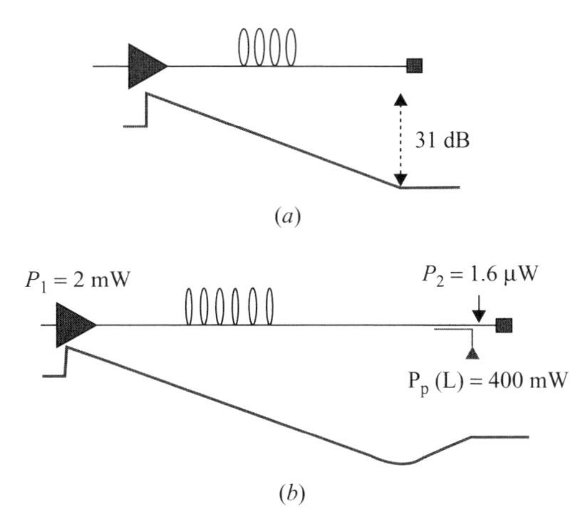

FIGURE 10.14 (*a*) Using an EDFA it is possible to have a certain link length before amplification would become necessary. (*b*) In case a longer span is required, Raman pumping in the backward direction can be used to increase the span length.

Raman amplification is strongly polarization dependent; the gain for signal polarization parallel to the pump is about an order of magnitude larger than the gain of signal polarization that is perpendicular to the pump. This is usually overcome by using polarization-diversity pumps in which pumps of both polarizations are launched simultaneously through the fiber.

There are some specific issues associated with Raman amplifiers. Although amplification using the Raman effect was demonstrated quite some time ago, due to nonavailability of compact high-power sources and small Raman gains it did not

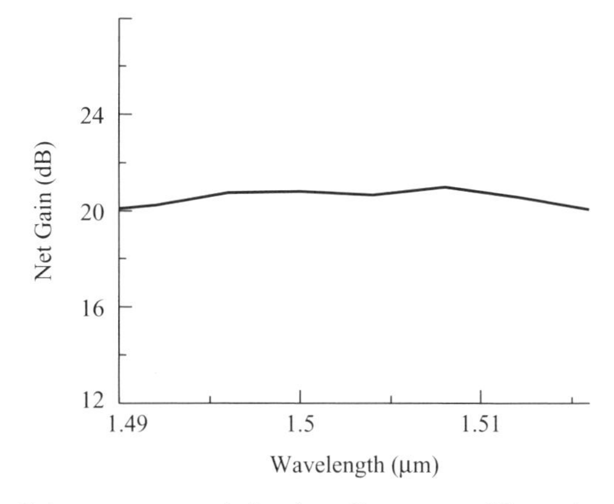

FIGURE 10.15 Gain spectrum variation in a Raman amplifier using a segmented clad design. Simulations correspond to a forward Raman pump power of 200 mW and a backward Raman backward pump power of 300 mW.

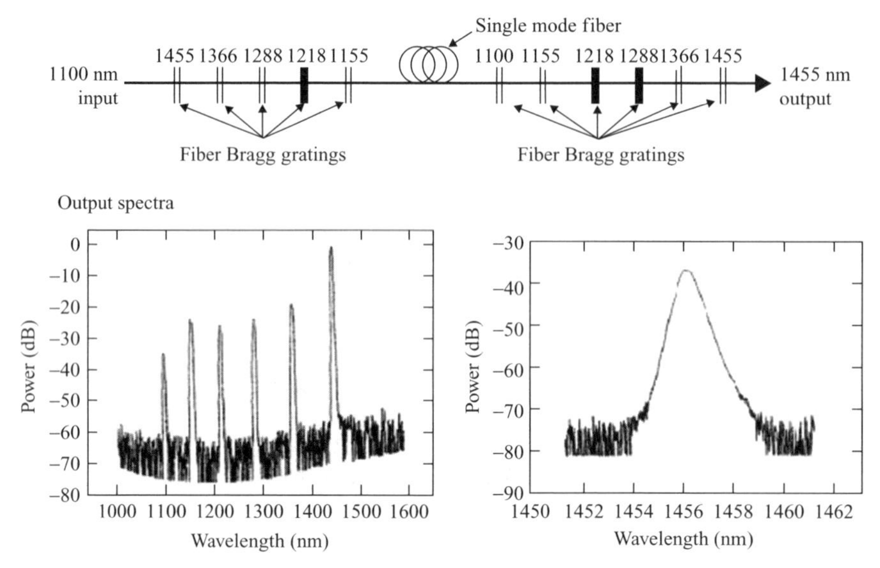

FIGURE 10.16 Cascaded Raman laser; output can be generated anywhere from 1100 to 1600 nm. (Adapted from Rottwitt, 2005.)

become important commercially. Recent developments in compact high-power semiconductor laser sources capable of giving 500 mW of power and optical fibers with high-Raman-gain coefficients have revived interest in these amplifiers. Since amplification is taking place over a long length of the fiber, additional noise due to the phenomenon of double Rayleigh scattering (DRS) is also generated. Signals propagating in the forward direction suffer Rayleigh scattering and generate power in the backward direction. Backward-propagating signals can undergo further Rayleigh scattering (double Rayleigh scattering) and generate power in the forward direction, adding to the noise. These signals use the same pump power to get amplified and constitute DRS noise. This noise becomes important for reasonably large pump powers and long interaction lengths when the gain becomes large. For nice reviews of fiber Raman amplifiers, readers are referred to Islam (2002) and Bromage (2004).

The Raman effect can be used to build a cascaded fiber Raman laser (Fig. 10.16). The vertical bars represent fiber Bragg gratings (FBGs), which are strongly reflecting at the wavelengths written on the top (see Chapter 11 for a brief account of FBGs). Thus, the high-power input wavelength of 1100 nm ($\approx 9091\ \text{cm}^{-1}$) produces Raman-scattered light at 1155 nm ($\approx 8658\ \text{cm}^{-1}$, implying a Raman shift of about $433\ \text{cm}^{-1}$), producing amplification at this wavelength. The FBGs reflecting at 1155 nm act as mirrors for this wavelength, and this results in laser oscillation at 1155 nm. Now, this strong light at 1155 nm ($\approx 8658\ \text{cm}^{-1}$) beam gets Raman scattered and produces light at 1218 nm ($\approx 8210\ \text{cm}^{-1}$, implying a Raman shift of about $448\ \text{cm}^{-1}$), which now resonates between the two FBGs with a peak reflectivity at 1218 nm. By using this cascaded process, laser output can be generated anywhere from 1100 to 1600 nm (Fig. 10.16).

Very recently, an ultralong 75-km laser has been realized using Raman scattering. Two pump lasers at 1365 nm are used at either end of the 75-km-long fiber, which has two FBGs with high reflectivities (about 98%) at a wavelength of 1455 nm; this wavelength corresponds to Raman-scattered light for the wavelength 1365 nm. The two gratings thus form the laser cavity, and the 75-km-long fiber forms the laser, with Raman scattering providing amplification.

Raman fiber amplifiers have helped expand the operating wavelength band of optical amplifiers to the entire low-loss window of transmission of silica-based optical fibers. Combining Raman fiber amplifiers with EDFAs to form hybrid amplifiers can lead to additional benefits in terms of better noise figure and extended reach of optical fiber communication systems.

Example 10.1 As an example of how Raman amplifiers can be used to increase the span length of a fiber optic link, consider a 16-channel system with an EDFA booster amplifier with an output power of 15 dBm ($\simeq$ 32 mW). This implies that each channel has a power of 2 mW. If the receiver sensitivity is -28 dBm ($\simeq$ 1.6 μW), then assuming a fiber loss of 0.25 dB/km, the maximum length of the span would be 124 km, corresponding to a loss of 31 dB. If we now launch a 400-mW pump propagating in the backward direction, this would result in a Raman gain of 15 dB. Thus, in the presence of the pump, the length of the fiber after which the signal power would drop to 1.6 μW would be such that

$$-0.25(\text{dB/km}) \times L(\text{km}) + 15(\text{dB}) = -31(\text{dB})$$

which gives $L = 184$ km. Thus, Raman amplifiers can be used to increase span length.

CHAPTER 11

Fiber Bragg Gratings

11.1 INTRODUCTION

In a normal optical fiber, the refractive indices of the core and of the cladding do not change along the length of the fiber. In fact, any deviation from this leads to scattering of the propagating light and to additional loss in the optical fiber. On the other hand, if we induce a periodic modulation of refractive index along the length in the core of the optical fiber, such a device, referred to as an *optical fiber grating*, exhibits very interesting spectral properties and such fiber gratings are finding many applications in fiber optic communications and sensing.

There are two main types of fiber gratings: *short-period gratings* or *fiber Bragg gratings* (FBGs) and *long-period gratings* (LPGs). In the case of FBGs, the spatial period of the modulation is about half a micrometer (i.e., the refractive index of the fiber core varies periodically along the fiber length and has the same value after every half a micrometer or so). Such a grating couples light at a specific wavelength propagating in the forward direction in the optical fiber to light propagating in the reverse direction in the fiber, thus acting like a mirror (Fig. 11.1). LPGs, which have spatial periods of refractive index modulation of a few hundreds of micrometers, couple light at specific wavelengths propagating in the core in the forward direction into the cladding, causing it to be lost from guidance (Fig. 11.2). Due to the periodic nature of the coupling process, FBGs and LPGs both have very interesting wavelength characteristics and thus find many uses in telecommunications and sensing.

11.2 REFLECTION OF LIGHT

When light encounters a dielectric of a different refractive index, it suffers from partial reflection. For example, when light falls normally on a glass surface (having a refractive index of 1.5), about 4% of the incident light is reflected. Similarly, silicon (which is used extensively in microelectronic circuits) has a refractive index of about 3.5 and thus reflects about 31% of the incident light, thus exhibiting a mirrorlike surface. The amount of reflection depends, of course, on the angle of incidence; as

Fiber Optic Essentials, By K. Thyagarajan and Ajoy Ghatak

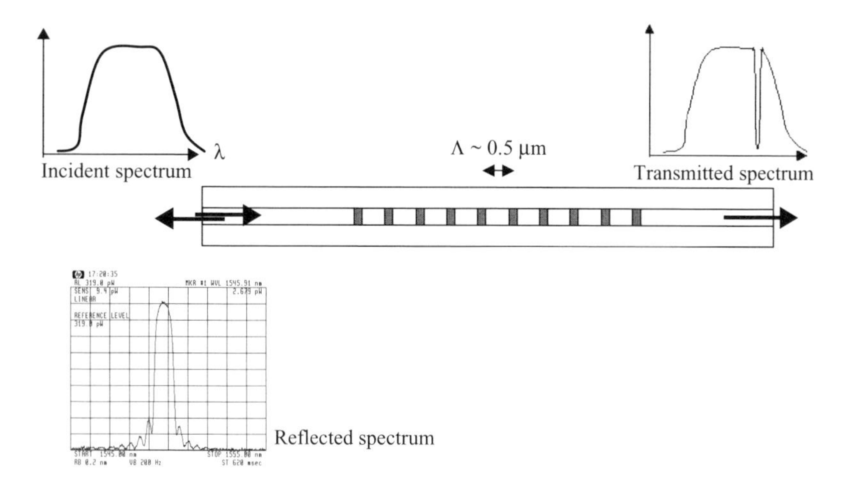

FIGURE 11.1 An FBG reflects light of a specific wavelength and thus acts like a wavelength-selective reflector. The period of the refractive index variation in the core is about half a micrometer.

the incident light reaches grazing angles of incidence, the reflectivity increases to almost 100%; this is easy to see in a glass window.

Optical instruments such as cameras, microscopes, and telescopes consist of many glass lenses, and if each surface reflects 4% of the incident light, this would result in a significant loss of light. Thus, a camera having 10 lens elements will transmit only 44% of the incident light, leading to a heavy loss of light due to reflection from the lens surfaces. It is possible to reduce the reflection from each layer by having a thin-film dielectric coating of appropriate thickness. If we consider a glass surface coated with a dielectric film, the incident light undergoes reflection from each of the two layers (Fig. 11.3), and if the thickness of the layer is chosen appropriately,

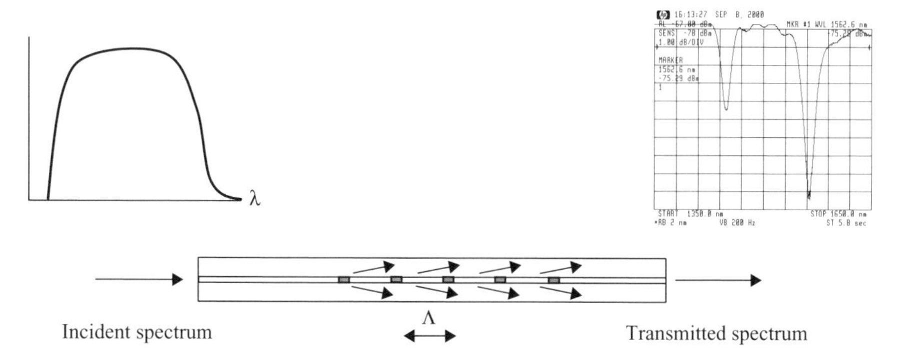

FIGURE 11.2 LPG-coupled light propagating in the core to the cladding, where it is lost, thus inducing a wavelength-dependent loss in the transmission. The period of refractive index variation in the core is a few hundred micrometers.

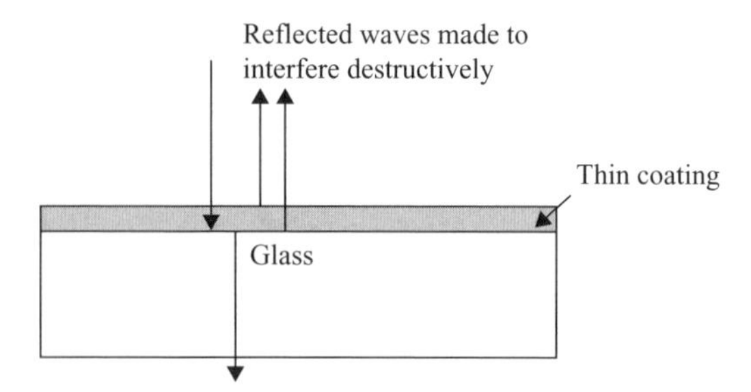

FIGURE 11.3 Reflection from a glass surface can be reduced by coating it with a transparent film; the two reflections occurring from the top and bottom surfaces of the film are made to interfere destructively so that there is no reflection.

the two reflected waves can be made to interfere with each other destructively, resulting in reduced reflection. These films are referred to as *anti reflection coatings*. Typically, the layers are made of magnesium fluoride or aluminum oxide and are about 0.1 μm thick (1 μm is one-thousandth of a millimeter). Destructive interference takes place at only one design wavelength, while at nearby wavelengths there is still some residual reflection. Hence, coated lens surfaces have a typical color that can be seen on spectacles, cameras, and binoculars, for example.

Instead of a single thin film, if we consider a multiple-layer structure consisting of a number of alternate layers of higher and lower refractive indices (Fig. 11.4), when a light wave enters the multiple-layer structure, it undergoes minute reflections from every interface. If all the individual reflections are in phase (i.e., the crest of each wave matches the crest of every other wave, and similarly for the troughs), the medium will strongly reflect the incident wave. If the reflected waves are not in phase, the net reflection would be weak. Since this phenomenon depends on wavelength, the overall reflection from such a multiple-layer medium would be very strongly wavelength dependent.

The wavelength of maximum reflectivity, referred to as the *Bragg wavelength* (λ_B), is given by

$$\lambda_B = 2n_0\Lambda \qquad (11.1)$$

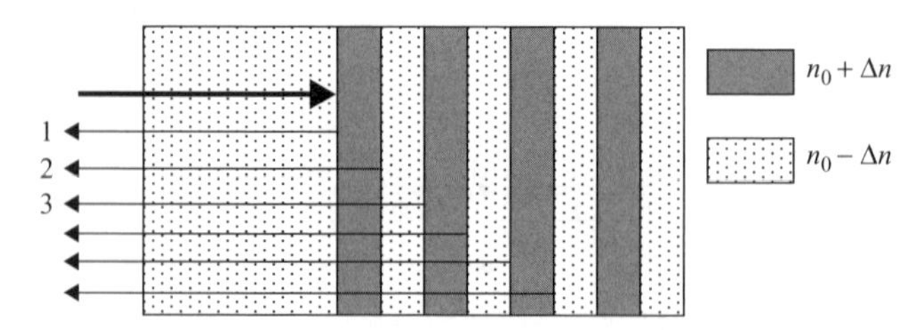

FIGURE 11.4 Reflections from a multilayer structure. If all the reflections from each layer interfere constructively, there would be a very strong reflection. Since this condition is wavelength dependent, such a structure would reflect strongly at certain wavelengths only.

FIGURE 11.5 Iridescent colors of a butterfly, due to interference between reflections from different layers.

where n_0 is the average refractive index of the medium. Equation (11.1) is referred to as the *Bragg condition* (reminiscent of X-ray diffraction from atomic planes in crystals), and the specific wavelength λ_B satisfying Eq. (11.1) is the *Bragg wavelength*. As the wavelength deviates from the Bragg wavelength, the waves reflected from the layers will not add constructively, and thus the reflection would drop down. Such multilayer coatings are used to achieve high-reflectivity mirrors for applications in lasers. Unlike normal mirrors, which are realized using metal coatings, they are made only of dielectric media, and exhibit very low reflection losses and can also be made highly wavelength selective.

Nature produces beautiful colors using this phenomenon of interference. Morpho butterflies and many varieties of beetles show brilliant colors that change with the orientation of the view (Fig. 11.5). Such colors are referred to as *iridescent colors*. The wings of these insects have naturally occurring multiple stacks of layers; interference of light reflected from the multiple layers is responsible for their colors. Paints are also available that use the production of color by interference; these contain tiny transparent flakes of mica coated with metal oxides (iron oxide or titanium dioxide) on all sides. Interference of light waves reflected from the various interfaces between the paint and the mica produces iridescence.

11.3 FIBER BRAGG GRATING

We now consider periodic modulation of the refractive index within the *core of a single-mode optical fiber*. Since light propagates with an effective refractive index given by n_{eff} within the fiber, instead of Eq. (11.1), we will have the equation

$$\lambda_B = 2n_{\text{eff}}\Lambda \tag{11.2}$$

The period and effective index of the fiber determine the Bragg wavelength at which the reflection would be maximum. When light at the Bragg wavelength is incident in the optical fiber, as it propagates through the periodic variation, the forward-propagating light beam will undergo strong reflection, leading to a propagating light in the backward direction through the fiber. Such a periodic variation within the core of the optical fiber is referred to as a *fiber Bragg grating* (FBG). When broadband light or a set of wavelengths are incident on an FBG, only the wavelength corresponding to the Bragg wavelength will get strongly reflected; the other wavelengths just get transmitted to the output (Fig. 11.1). The wavelength of maximum reflectivity, also referred to as the *Bragg wavelength*, is proportional to the period of the index modulation and to the refractive index of the fiber.

The peak reflectivity R and bandwidth $\Delta\lambda$ (spectral width over which the reflectivity is high) of a fiber Bragg grating of length L are given approximately by

$$R = \tanh^2 \kappa L \tag{11.3}$$

and

$$\Delta\lambda \approx \frac{\lambda_B^2}{n_{\text{eff}} L}\left(1 + \frac{\kappa^2 L^2}{\pi^2}\right)^{1/2} \tag{11.4}$$

where

$$\kappa \approx \frac{\pi \Delta n I}{\lambda_B} \tag{11.5}$$

Here Δn is the peak refractive index change within the grating and I is given by

$$I \approx 1 - \exp\left(-\frac{2a^2}{w^2}\right) \tag{11.6}$$

where a is the fiber core radius and w represents the Gaussian spot size of the fundamental mode; the Gaussian spot size is discussed briefly in Chapter 4.

Using Eqs. (11.3) to (11.6), it is possible to obtain the properties of a grating with a certain period and length or to estimate the grating parameters required to achieve a desired peak reflectivity and bandwidth knowing the fiber properties. As an example, let us consider a grating made in a fiber with $a = 3$ μm, NA $= 0.1$, and $n_2 = 1.45$. The fiber is single-moded at 850 nm, and the fundamental mode has an effective index of 1.4517. If we wish to strongly reflect a wavelength of 850 nm, the corresponding spatial period required would be

$$\Lambda = \frac{\lambda_B}{2n_{\text{eff}}} \approx 0.293\ \mu\text{m}$$

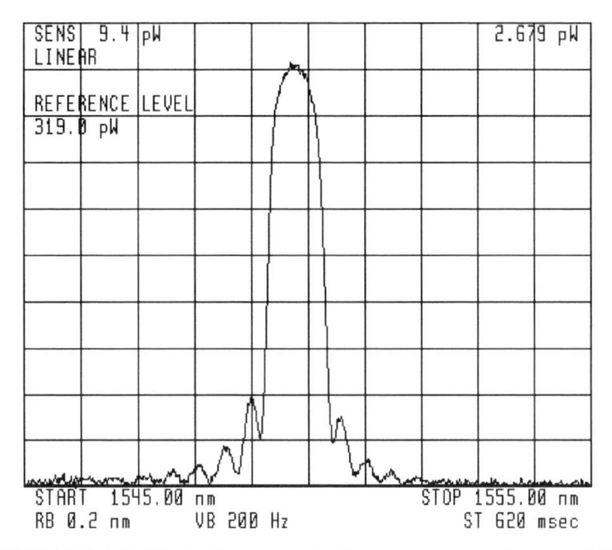

FIGURE 11.6 Typical measured reflection spectrum from an FBG.

Note that the required spatial period is less than the optical wavelength. If the grating has a length of 1 cm, the required value of κ for 20% peak reflectivity would be 0.48 cm^{-1} which gives a Δn value of 1.3×10^{-5}. The corresponding bandwidth is $\Delta\lambda \approx 0.05$ nm. We have assumed $I \approx 1$. Note also the extremely small bandwidth of the grating.

From Eqs. (11.3) and (11.4) it also follows that one can design gratings having the same peak reflectivity but different bandwidths by appropriate choice of the peak index modulation and grating length. A given peak reflectivity implies a value of κL (i.e., $\Delta n\, L$). The bandwidth can be increased or decreased by increasing or decreasing the grating length while keeping the product $\Delta n\, L$ constant.

Figure 11.6 shows a measured reflection spectrum of an FBG. The peak wavelength of reflection is about 1549.8 nm, and the wavelength region of high reflectivity is about 1 nm wide, showing the strong wavelength selectivity of such gratings. Wavelengths not falling within this bandwidth get transmitted through the grating. Figure 11.7 shows the measured reflection spectrum of four gratings written in one fiber. The peak wavelengths of the four gratings are 1534.275, 1543.192, 1552.121, and 1561.090 nm, and the length of each grating is 15 mm.

Example 11.1 Let us consider a fiber with a fundamental mode effective index of 1.45. If we wish to design an FBG reflecting at a wavelength of 1550 nm, the period of the grating required will be about 0.534 μm. Note the very small period. If the grating has a length of 2 cm and the refractive index modulation is 8×10^{-5}, the peak reflectivity would be about 99.4% and the bandwidth would be 0.12 nm. Note the extremely small bandwidth of the grating. Thus, FBGs are very narrowband reflectors.

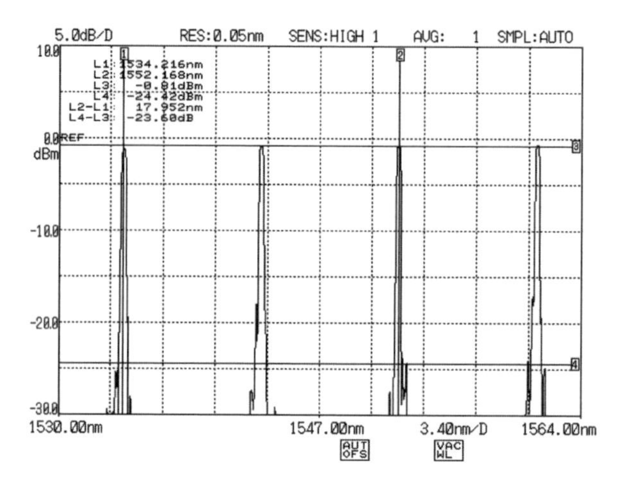

FIGURE 11.7 Reflection spectra of four FBGs fabricated on the same fiber. Peak reflectivities of the FBGs occur at 1534.275, 1543.192, 1552.121, 1561.090 nm; length of each grating, 15 mm; reflectivity, >90%; bandwidth, 0.27 to 0.29 nm; spacing between the gratings, 200 mm. (Courtesy of Dr. S. K. Bhadra, CGCRI, Kolkata, India).

Example 11.2 Since the wavelength corresponding to the peak reflectivity is directly proportional to the period of index modulation, a 0.1% change in the period would result in a 0.1% change in the wavelength of peak reflectivity. Thus, if in Example 11.1 the period is changed to about 0.539 μm, the wavelength of peak reflectivity would change to approximately 1551.55 nm.

11.4 SOME APPLICATIONS OF FBGs

Fiber Bragg gratings find many uses in telecommunication and sensing. These include applications in fiber grating sensors and add/drop multiplexers, and to provide external feedback for laser diode wavelength locking, dispersion compensation, and the like. Table 11.1 gives some important applications of FBGs. Here we discuss some of the applications.

Add/Drop Multiplexer

In many telecommunication networks carrying wavelength-division-multiplexed signals, it is required to add or drop a wavelength channel. In such a case we need to have devices that can choose the desired wavelength and filter it out of the link or let us add signal corresponding to the wavelength chosen. Figure 11.8 shows a typical configuration of an add/drop multiplexer based on an FBG. An FBG that reflects light at the wavelength that needs to be dropped is placed between two optical circulators. (The circulator is a device in which light incident from port 1 gets transmitted to port 2

TABLE 11.1 Applications of Fiber Gratings

Application	Grating Property	Typical Parameters
Laser wavelength stabilization	Narrowband reflector	$\Delta\lambda \sim 0.1$ to 1 nm, $R = 1$ to 100%
WDM add/drop filter	High-isolation reflector	$\Delta\lambda \sim 0.1$ to 1 nm, isolation > 50 dB
EDFA gain equalizer	Adjustable transmission spectrum	$\Delta\lambda \sim 30$ nm, loss = 0 to 10 dB
Dispersion compensation	Differential delay using chirped grating	$\Delta\lambda \sim 0.1$ to 15 nm, dispersion = 1600 ps/nm
Sensors	Sensitivity toward external perturbations	Bragg wavelength shifts, in picometers

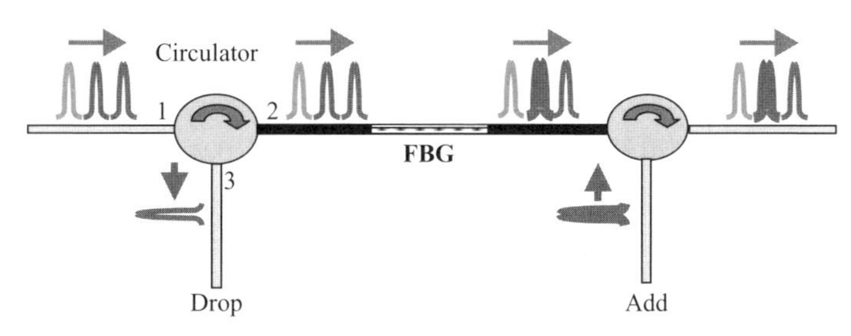

FIGURE 11.8 Add/drop multiplexer using FBGs.

and light incident from port 2 gets transmitted to port 3.) From among the incoming wavelength channels of the DWDM transmission system, the wavelength matching the FBG wavelength will get reflected and be routed to the dropped port. All other wavelength channels proceed forward along the link. At the same time, signal at the same wavelength can be added using the second circulator with the same grating reflecting the channel added to the link. Since an FBG reflects only the wavelength chosen and transmits all other channels, it is possible to add and drop multiple channels by using multiple FBGs at the desired wavelengths instead of a single FBG. The gratings used in add/drop multiplexers need to have high reflectivity so that there is no residual signal leading to crosstalk among the dropped and added channels.

Dispersion Compensation

It is well known that when pulses of light propagate through an optical fiber link, they suffer from pulse broadening due to the dependence of the velocity of propagation on the wavelength. This dispersion needs to be compensated in the link, and this can be achieved either through the use of dispersion-compensating fibers (Chapter 7) or with the help of chirped FBGs.

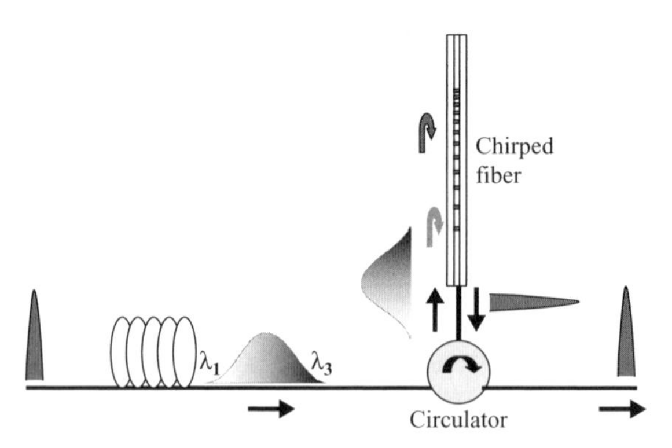

FIGURE 11.9 Dispersion compensation using chirped FBG in which the period of the grating changes along the length.

In a uniform FBG, the period of the refractive index modulation is constant along the length of the grating. If the period of the grating varies along its length, this is referred to as a *chirped fiber Bragg grating*. Since the period of the grating varies along the length in such chirped FBGs, the Bragg wavelength also varies along the position in the grating (Fig. 11.9). When light propagates through such a grating, different wavelength components present in the incident wave will get reflected at different positions along the grating; this will lead to different wavelength components having different time delays to return to the input end. By using an appropriately chirped FBG, one can indeed compensate for the differential delay of different wavelengths accumulated while propagating through an optical fiber link.

Let us consider propagation of a pulse of light through an optical fiber operating at a wavelength longer than the zero-dispersion wavelength of the fiber. This would correspond to, say, operating a G.652 fiber (having zero dispersion at 1310 nm) at a wavelength of 1550 nm. Thus, if we consider three wavelength components $\lambda_1 > \lambda_2 > \lambda_3$ within the pulse, we can see that due to dispersion in the fiber, wavelength λ_1 would suffer a larger delay than wavelength λ_2, which in turn will suffer a delay longer than λ_3 while propagating through the fiber. To compensate for this dispersion, we need to delay wavelength component λ_3 more than component λ_2, which in turn should suffer a delay more than at component λ_1. To achieve this, the chirped grating is designed so that wavelength λ_1 reflects from the near end of the grating, λ_2 reflects from a portion farther away, and λ_3 reflects from the far end, to compensate for the differential delay between all the wavelength components, thus leading to dispersion compensation.

The dispersion introduced by the grating of length L_g is given by the following formula:

$$\frac{d\tau}{d\lambda} = \frac{8\pi n_{\text{eff}}^2}{c\lambda_0^2}\frac{L_g^2}{F} \tag{11.7}$$

where F is the dimensionless chirp parameter, which determines the variation of grating period with length. Thus, to compensate for accumulated dispersion of DL_f

of the link fiber with dispersion coefficient D and length L_f, we require that

$$DL_f = -\frac{8\pi n_{\text{eff}}^2}{c\lambda_0^2}\frac{L_g^2}{F} \tag{11.8}$$

The frequency bandwidth over which compensation will take place is given by

$$\Delta\nu = \frac{c}{4\pi n_{\text{eff}}}\frac{F}{L_g} \tag{11.9}$$

If $\wedge_0$ is the average grating period and $\Delta\wedge$ is the difference in grating period between the two ends of the grating,

$$F = \frac{2\pi L_g \Delta\wedge}{\wedge_0^2}$$

As an example, let us consider a chirped grating of length 11 cm with the chirp parameter $F = 640$, operating at an average wavelength of 1550 nm with a fiber effective index of 1.45. Using Eqs. (11.8) and (11.9), we obtain for the dispersion of the grating 1380 ps/nm operating over a bandwidth of 0.61 nm. This grating can compensate for dispersion accumulated over a fiber with a dispersion coefficient of 17 ps/km·nm of length 81 km over a bandwidth of 0.61 nm, which is approximately equal to 76 GHz. It is interesting to note that the difference in grating period between the front and back ends of the 11-cm-long grating is only about 0.25 nm, while the average period of the grating is about 0.534 μm.

Chirped dispersion-compensating gratings are available commercially for compensation of accumulated dispersion of up to 80 km of G.652 fiber for up to 32 wavelength channels. Figure 11.10 shows a typical variation of group delay with

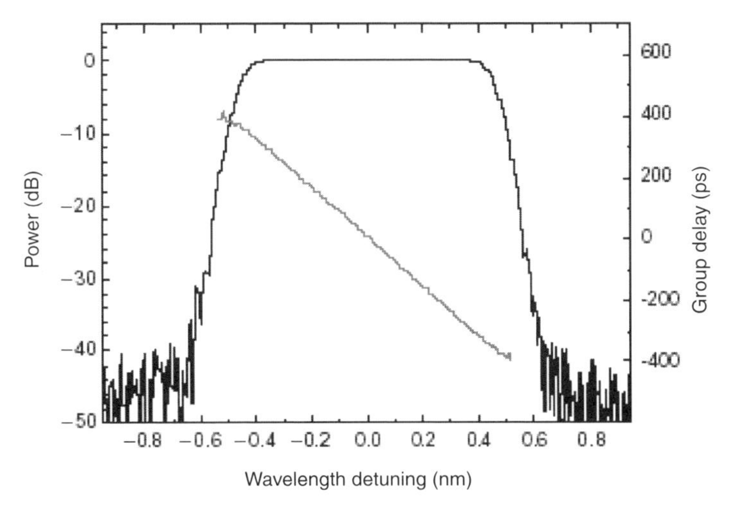

FIGURE 11.10 Variation of group delay and reflectivity of a typical commercial dispersion compensating grating. (Adapted from Sifam Fiber Optics Ltd., UK.)

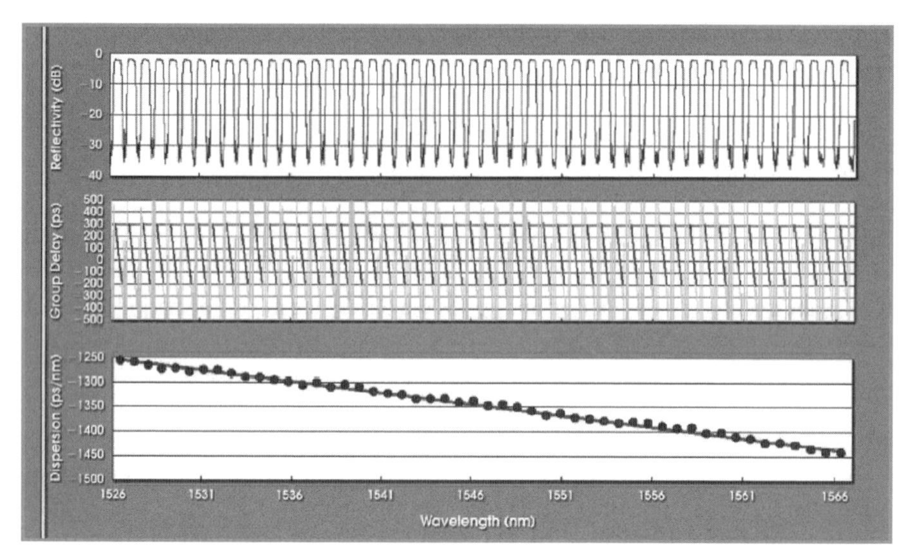

FIGURE 11.11 Dispersion-compensating FBG for 51 wavelength channels spaced 100 GHz apart. (Adapted from Guy and Painchaud, 2004.)

wavelength; the slope of this curve will give the dispersion. The grating can provide dispersion compensation between 700 and 1400 ps/nm. Unlike dispersion-compensating fibers, chirped FBGs provide the possibility of tweaking the required dispersion compensation, especially for 40-Gbps systems, where the margin of dispersion available is rather small. Also, by using nonlinearly chirped FBGs it has been shown that delay variation from −200 to −1200 ps is possible. For achieving dispersion-compensating gratings operating over a large band of wavelengths for WDM applications, various techniques have been used. Figure 11.11 shows dispersion-compensation possibilities using a phase-sampling technique showing compensation possibility for 51 channels spaced by 100 GHz. Polarization mode dispersion is another important effect that degrades pulse propagation and becomes important, especially at high bit rates. Compensation of PMD can be achieved using chirped FBGs fabricated in high-birefringence fibers.

FBG-Based Sensors

FBGs have a great potential for applications as sensors for sensing mechanical strain, temperature, acceleration, and so on. Since the Bragg wavelength (wavelength of maximum reflectivity) depends on both the refractive index of the fiber and the period of the grating, any external parameter that changes any of these would result in a change in the reflected wavelength. Thus, by measuring *changes in the reflected wavelength*, the external perturbations affecting the grating could be sensed. This is the basic principle of their application in sensing.

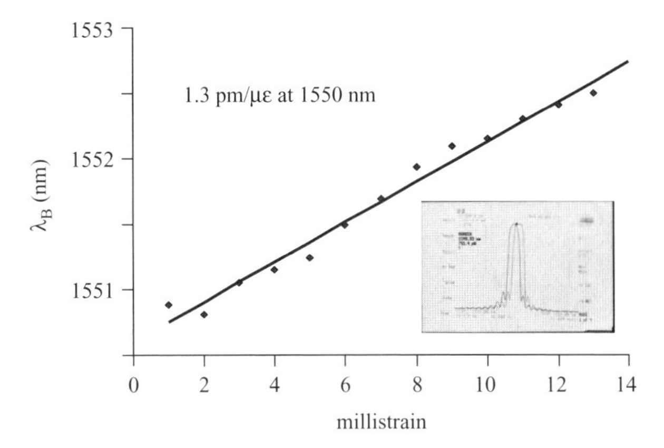

FIGURE 11.12 Variation measured in the peak wavelength of reflection with the strain in the fiber.

When a fiber grating is strained, say by pulling it, the grating suffers an elongation that leads to an increase in the period of the grating. At the same time, strain on the fiber leads to minute changes in the refractive index of the fiber through the *strain-optic effect*. Together, these effects lead to a change in the Bragg wavelength of the grating. Figure 11.12 shows a variation of Bragg wavelength measured as a function of longitudinal strain (pulling) of the grating. The grating shown has a strain sensitivity of 1.3 pm/$\mu\varepsilon$ (picometer per microstrain) at 1550 nm. This means that if a 1-m-long fiber is pulled by a length of 1 μm, this would result in a strain of 1 $\mu\varepsilon$, which would result in a wavelength shift of 1.3 pm of the reflected wavelength (1 pm is one-millionth of one-millionth of a meter, or one-billionth of a millimeter).

Since the refractive index as well as the period of the grating change with change in temperature, the temperature change of an FBG would also result in a change in the peak reflected wavelength. The temperature sensitivity of FBGs is typically about 6 pm/°C (i.e., a change of temperature of 1°C will result in a change in the reflected wavelength by 6 pm).

From the discussions above it can be seen that changes in the peak wavelength are indeed very small, and hence special techniques are needed for sensing such small changes. Since changes in reflected wavelength could be due to strain as well as to temperature changes, techniques to deconvolve (separate) changes in temperature and strain need to be implemented for precise sensing.

One of the important attributes of FBG sensors is that they can be multiplexed into a single fiber. Figure 11.13 shows a typical arrangement of a multiplexed sensing arrangement in which a number of FBGs with different Bragg wavelengths are fabricated at different points along the length of a single-mode fiber. Light from a broadband source is coupled into the fiber, and light at different wavelengths is reflected from the individual gratings and analyzed by the detection circuit. The wavelengths of the FBGs are chosen such that they do not overlap each other and fall within the band of the source. By measuring changes in the Bragg wavelength

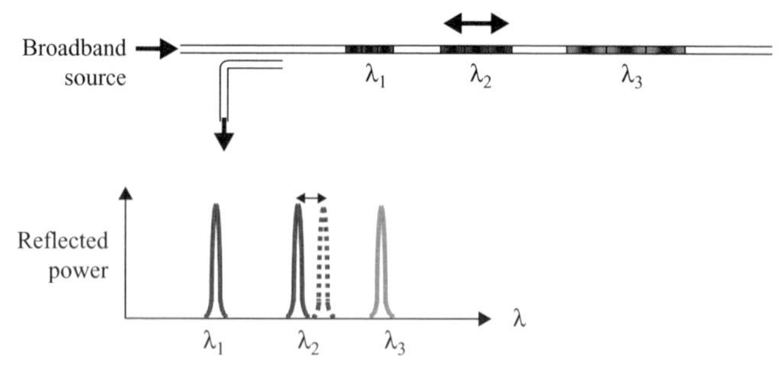

FIGURE 11.13 Multipoint sensor based on FBGs reflecting at various wavelengths. By monitoring the spectrum reflected from the fiber, the location of the strain and its magnitude can be estimated.

of individual FBGs, the strains or temperature changes at each FBG location can be measured independently.

Currently, there is intense activity in achieving different multiplexing geometries with high-sensitivity measurements. Preliminary experiments are under way to use such sensors in real-world environments such as in bridges and other civil structures. Figure 11.14 shows a 350-m-long five span bridge in Norway wherein 32 FBG sensors were laid for strain monitoring. Figure 11.15 shows the response of the grating sensor as a 50-ton lorry is driven across the bridge at 30 km/h. Measurements carried out using FBG sensors were found to be consistent with measurements using other conventional sensors.

FIGURE 11.14 FBG sensors on a bridge in Norway. (Adapted from Gebremicheal et al., 2005. Copyright © 2005 IEEE.)

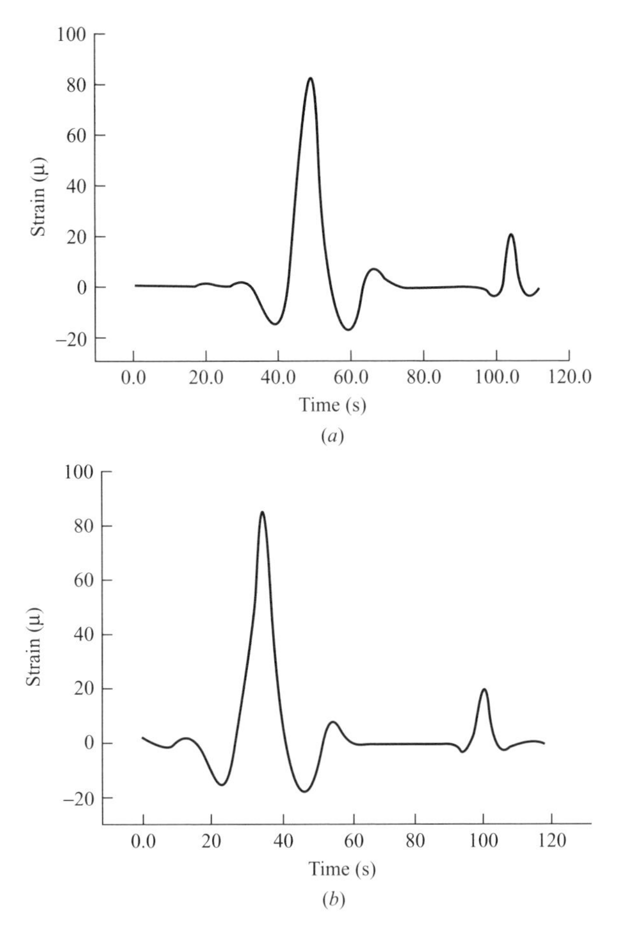

FIGURE 11.15 Response of the grating sensor as a 50-ton lorry is driven at 30 km/h across the bridge in Fig. 11.14. (Adapted from Gebremicheal et al., 2005. Copyright © 2005 IEEE.)

Among the various issues in connection with FBG sensors is the problem of separating the changes brought about by temperature and strain. There is considerable work going on to achieve this. Fiber optic sensing technology is advancing rapidly, and FBGs are now being tested for structural monitoring. In the future it may be possible to multiplex perhaps 100 sensors on a single fiber, which gives enormous capability to structural monitoring.

As compared to other types of fiber optic sensors, FBGs have some distinct advantages:

- They are immune to intensity fluctuations of the source since the sensor signal is in the form of a shift in peak reflection wavelength rather than changes in the intensity of the light.
- The Bragg wavelength is almost a linear function of strain or temperature.

- Fiber grating sensors can be written directly on the fiber and are mass producible.
- Multiplexing of FBGs is relatively easy, and thus they can be configured into quasi-distributed sensor assemblies.

FBG sensors are especially attractive for quasi-distributed sensing applications wherein FBGs with different center wavelengths can be used to sense signals at different points along the same fiber. Since different FBGs reflect light at different wavelengths, signals coming from different points along the fiber can easily be differentiated.

Example 11.3 For a Bragg wavelength of 1550 nm, a change of 0.1 μm in the length of a 1-cm grating corresponds to a strain of $0.1\ (\mu\text{m})/10^4\ (\mu\text{m}) = 10^{-5} = 10\ \mu\varepsilon$. The corresponding shift in the Bragg wavelength is given by $1.3 \times 10 = 13$ pm $\sim$ 0.013 nm, which is indeed a very small shift.

Example 11.4 If a mass of 250 g is hung on an FBG, for a 125-mm-diameter fiber, the stress will be about $2 \times 10^8\ \text{N m}^{-2}$. The Young's modulus ($Y$) of silica is about 72 GPa ($= 72 \times 10^9\ \text{N/m}^{-2}$). Thus, the corresponding strain will be about 2.8 millistrain. This strain will result in a shift of Bragg wavelength of $1.3 \times 2.8 \simeq 3.6$ nm.

11.5 FABRICATION OF FBGs

The fabrication of FBGs in optical fibers is based on an effect called *photosensitivity*, in which the refractive index of the germanium-doped fiber core can be increased permanently by exposure to ultraviolet light, typically at a wavelength of 248 nm. Exposure to ultraviolet light causes refractive index changes of about 0.001. To create such a change, excimer lasers or frequency-doubled argon ion lasers are employed. A periodic refractive index modulation can be produced in a fiber core by exposing the fiber to an interference pattern formed by two ultraviolet (UV) beams (Fig. 11.16). In regions of constructive interference between the beams, the UV intensity would be high, which leads to a local increase in refractive index. At the same time, in regions

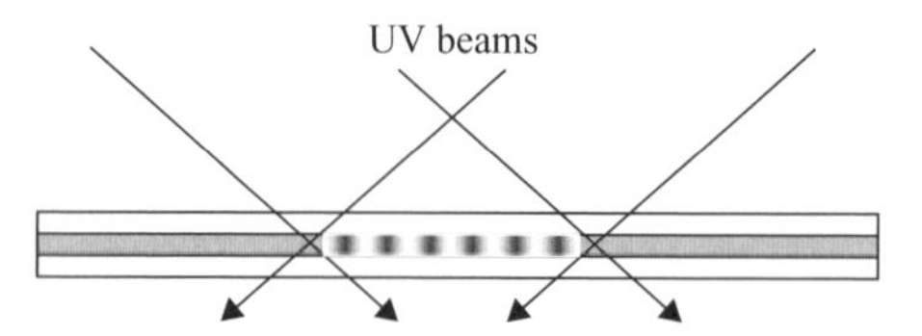

FIGURE 11.16 Fabrication of FBGs using interference between two ultraviolet laser beams; at the points of high intensity the refractive index of the fiber changes permanently, leading to formation of a grating.

of destructive interference, there is no change in the refractive index. Therefore, such an exposure would lead to a periodic modulation of the refractive index of the core of the fiber. The period of the modulation could be varied by varying the angle between the interfering beams. The interfering beams could also be produced using phase masks.

11.6 LONG-PERIOD GRATINGS

Long-period gratings (LPGs) are periodic perturbations along the length of the fiber with periods greater than 100 μm which induce coupling between light propagating in the core of the fiber to light propagating in the cladding and in the same direction. Since light propagating in the cladding is lossy, the coupled light gets lost from the fiber. The coupling process being wavelength selective, these gratings act as wavelength-dependent loss components. This makes them attractive candidates for applications in wavelength filters with specific application in gain flattening of erbium-doped fiber amplifiers, band-rejection filters, WDM isolation filters, or as polarization-filtering components and sensors, for example.

Figure 11.17 shows the transmission spectrum measured for an LPG fabricated in a single-mode optical fiber. The grating period is 463 μm and consists of 40 periods of index perturbation. One can see multiple dips in the transmission spectrum, corresponding to the coupling of the light propagating in the core to light propagating along different directions in the cladding (to different cladding modes). The corresponding refractive index perturbation is about 3×10^{-4}. The positions, depths, and widths of the loss dips in the transmission spectrum can be controlled by an appropriate choice of the period, length, and index modulation of the grating.

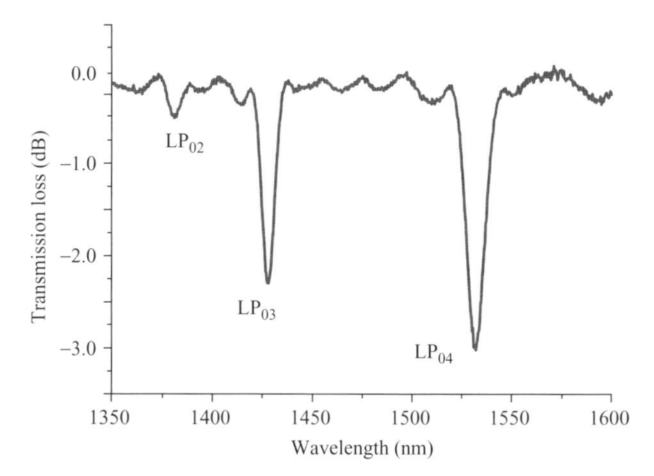

FIGURE 11.17 Measured transmission spectrum of a long-period grating fabricated in a single-mode optical fiber. Each dip corresponds to coupling of light in the core to a specific cladding mode. (Adapted from Palai et al., 2001.)

11.7 SOME APPLICATIONS OF LPGs

Some of the many applications of LPGs are discussed next.

EDFA Gain Flattening

As discussed in Chapter 9, the gain spectrum of an EDFA is not flat, which causes problems in their use in a communication system. Flattening of the gain spectrum is possible by using appropriate external wavelength filters, which compensate for the gain variation by having a transmission spectrum that is just the inverse of the gain spectrum. Since LPGs can be designed to have different transmission spectra, they find applications as gain-flattening filters. Figure 11.18*a* shows the transmissions profile of a LPG designed for gain flattening of an EDFA, and Fig 11.18*b* shows the

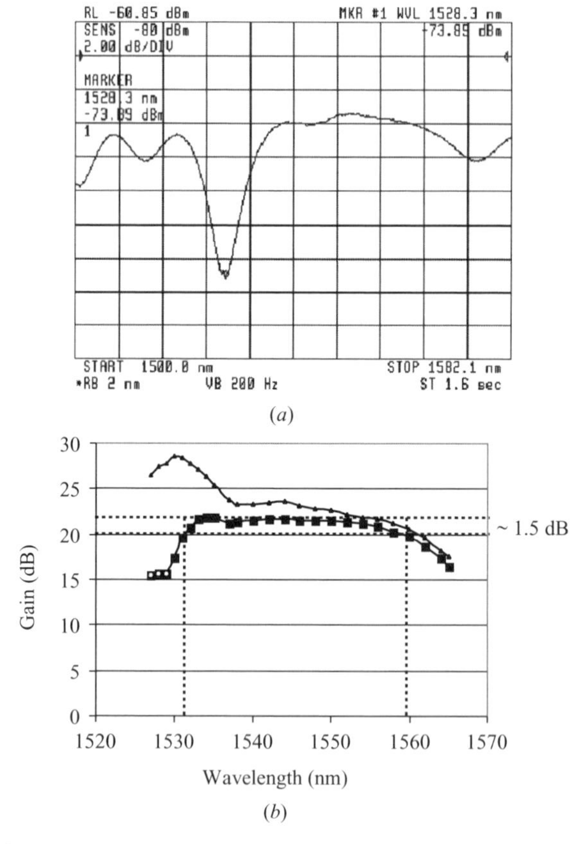

FIGURE 11.18 (*a*) Transmission spectrum of an LPG specially designed for gain flattening of EDFAs; (*b*) gain spectra corresponding to an EDFA with and without a gain-flattening filter.

gain spectrum of an EDFA with and without a filter. As can be seen, the LPG filter flattens the gain to within ±0.75 dB.

Mode Conversion

There are many applications wherein coupling of power among different guided modes propagating in the same direction in the fiber may be desired. Thus, in dispersion compensators using higher-order modes, we need to couple light from the fundamental mode to a higher-order mode that is also guided within the core. LPGs could be used to achieve such mode converters. Knowing the propagation characteristics of the two modes, LPGs of appropriate period could be designed which can achieve efficient mode conversion (see, e.g., Ramachandran, 2005).

Sensing

The transmission characteristics of LPGs are very sensitive to strain, temperature, bend, and so on. The temperature and strain sensitivity of an LPG can be as high as 3.4 nm/°C and −33.6 nm/mε. Hence, such sensors could be used in conjunction with FBGs to realize sensitive sensor configurations. Since in an LPG, light propagating in the core is coupled into light propagating in the cladding, any change in the medium surrounding the fiber cladding could influence the power transmitted through the fiber. This can be used to realize very sensitive refractive index sensors. They find special applications in the area of chemical and biological sensors. The sensitivity of these sensors can be enhanced by coating the cladding in a fiber region that has LPG with materials such as polymers and metals.

11.8 CONCLUSIONS

Fiber Bragg gratings and long-period fiber gratings have interesting spectral characteristics and are finding applications in many devices, such as add/drop multiplexers, wavelength lockers, dispersion compensators, sensors, and optical amplifier gain flatteners. By tailoring the grating properties and the fiber properties on which the gratings are fabricated, gratings with different spectral characteristics can be achieved.

CHAPTER 12

Fiber Optic Components

12.1 INTRODUCTION

Optical fiber components can be broadly classified as passive and active. Electrical powering is not required for *passive components*, which include multiplexers and demultiplexers, fixed-wavelength filters, interleavers, fixed optical add/drop multiplexers, dispersion compensators, couplers, splitters, pump combiners, isolators, and circulators. *Active components* include optical amplifiers, tunable wavelength filters, tunable dispersion compensators, wavelength converters, optical switches, external modulators, transmitters, and receivers. Optical sources (laser diodes) at different wavelengths, optical detectors, and optical fiber form the major components in any fiber optic communication system. Besides these, other components are required to amplify the signals in the optical domain and to compensate for accumulated dispersion. Many of these also contain fiber optic components such as optical taps, wavelength-division multiplexers and demultiplexers, and optical isolators.
Various technologies have been developed to achieve many of these functions:

1. *Fiber-based:* includes, for example, optical fiber amplifiers, fiber dispersion compensators, fiber Bragg gratings, and fused fiber components.
2. *Integrated optics–based:* includes, for example, arrayed waveguide devices, semiconductor optical amplifiers, and high-speed modulators.
3. *Microoptics and thin film–based:* includes, for example, filters, and multiplexers.

Various fiber optic components are listed and compared with corresponding bulk optical devices in Table 12.1.

We discussed optical amplifiers and dispersion compensators based on fibers and fiber Bragg grating–based dispersion compensators and wavelength-selective reflectors earlier. In this chapter we discuss other important components that are used in fiber optic communication systems. For a recent review of various fiber optic components, see, e.g., Pal (2006).

Fiber Optic Essentials, By K. Thyagarajan and Ajoy Ghatak

TABLE 12.1 Some All-Fiber Components Replacing Bulk Optical Components

Component	Bulk Device	All-Fiber Device
Beamsplitter		
Dichroic splitter	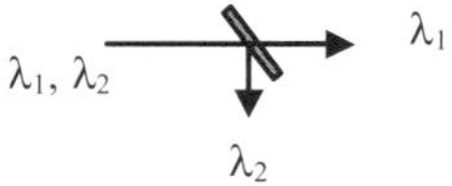 λ_1, λ_2 → λ_1; λ_2	 λ_1, λ_2 → λ_1; λ_2
Sagnac interferometer (see Chapter 14 for an application of this in a fiber optic gyroscope)	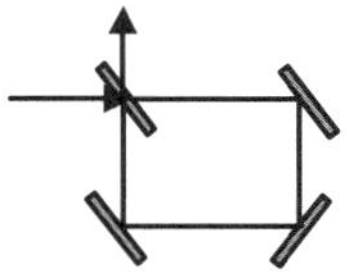	
Mach–Zehnder interferometer	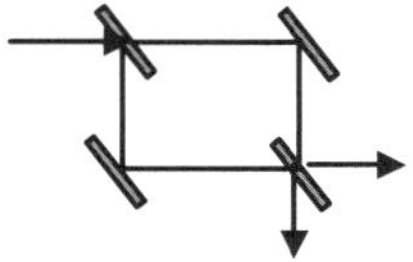	
Wavelength-selective mirror		
Polarization controllers	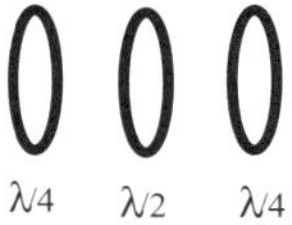 λ/4 λ/2 λ/4	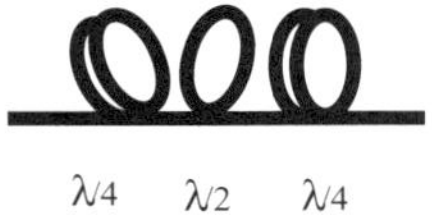 λ/4 λ/2 λ/4

12.2 FIBER OPTIC CONNECTORS

In many situations we need to have a demountable connection between fibers, between a fiber and a source or a detector, or between different fiber optic instruments. Demountable connections are achieved using fibers with connectorized terminations. Different designs of connectors have been developed over many years with the aim of reducing the loss or for ease of connection. The connector should ensure that the fiber core is aligned with the laser or the detector or the fiber to which it is being connected. Due to the smallness of the core of the fiber, connectors are high-precision devices with tolerances of less than 1 μm. Connector losses between two fibers will be minimized if the fibers are identical and are perfectly aligned and touch each other. Any gap between the two fibers will lead to a small loss but also to reflections from the fiber ends. This reflection can lead to problems, especially when they enter a laser diode. Hence, back reflection loss should be minimized in a connector. It is possible to reduce back reflection by using angle-polished end faces, since reflection from the ends do not couple back into the fiber provided that the angle of the polish is chosen appropriately. In a connector the fiber ends are polished and fixed. Any dirt on the fiber end can lead to losses when the connector is used.

Each connector consists of four major components: the ferrule, the connector body, the cable, and the coupling device. The ferrule is a long, thin cylindrical device, usually made of metal or ceramic, with a hole drilled through its center. The hole diameter is slightly larger than the cladding diameter so the fiber can be positioned snugly within the ferrule. Connectors with different ferrule shapes or finishes are available. Most currently used ferrules allow for physical contact between the two fibers being connected. Better physical contact is achieved by using a convex surface for the ferrules.

The connector body is usually made out of metal or plastic and holds the ferrule that carries the fiber. The ferrule extends beyond the connector body so as to be able to connect to another connector. The cable jacket and strength members are usually bonded or crimped to the connector body. The cable is attached to the body of the connector and acts as the point of entry for the fiber. The coupling device helps in the alignment of two fibers. Figure 12.1 shows a picture with different standard connectors: ST (straight terminus), SC (subscriber connector), FC (fiber connector), and LC (lucent connector). The connector losses are typically 0.15 to 0.3 dB, and the return loss (which is the ratio of the power reflected to the power in the forward direction) is about 55 dB.

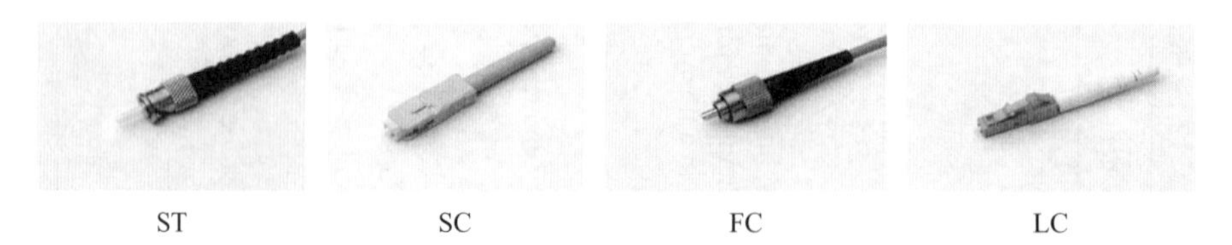

ST SC FC LC

FIGURE 12.1 Different common types of connectors in use. ST, straight terminus; SC, subscriber connector; FC, fiber connector; LC, lucent connector. (Adapted from http://www.thefoa.org/tech/ connID.htm.)

FIGURE 12.2 Typical fiber optic patch cord.

12.3 FIBER OPTIC PATCH CORDS

Fiber optic patch cords are the simplest fiber optic elements, consisting of a short length of optical fiber with a connector on either end (Fig. 12.2). Since they are used to connect various components and instruments in a fiber optic system, their characteristics in terms of loss and aging determine the overall performance of the system. In principle, when two patch cords are connected, if the fibers are identical, it should result in almost zero loss. In actual practice the loss may not be very small since the fiber may not be completely concentric with the connector center, there could be dust at the tip of the connector, or there could be misalignments when two patch cords are mated. Patch cords with different types of fibers and different connector types are available. Typical insertion loss of patch cords are about 0.4 dB, with a return loss of better than 50 dB.

12.4 FIBER OPTIC COUPLERS

In fiber optic communication systems, it is often necessary to tap a small amount of power from the signal. It may also be necessary to split the signal into two (or more) parts so that the same signal can reach two (or more) destinations. All this can be achieved by means of a coupler, which is essentially a fiber optic beamsplitter and is one of the most important inline fiber components. The schematic of a typical fiber optic directional coupler is shown in Fig. 12.3. The device works on the basic fact that even though light is confined to propagate along the core of the fiber, a small fraction of the light extends beyond the core–cladding interface, which of course propagates along with the light within the core at the same speed. Thus, when two fiber cores are brought sufficiently close (separation of the order of a micrometer or so) to each other laterally, periodic exchange of power takes place from one fiber to the other

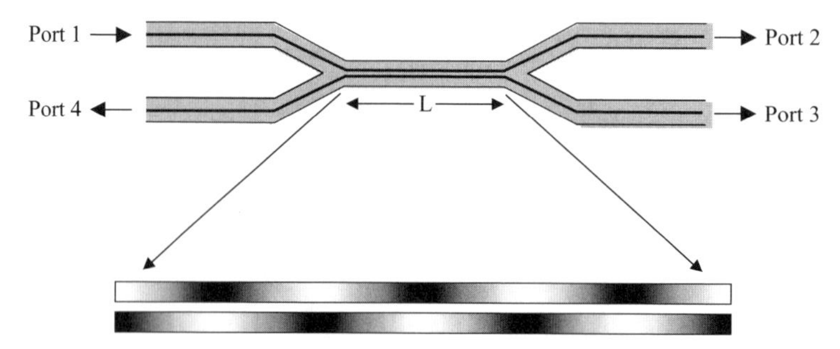

FIGURE 12.3 Directional coupler consisting of two fibers whose cores are brought close to each other. Due to interaction between the fibers, there is a periodic exchange of power between the two fibers, as shown in the lower part of the figure.

(Fig. 12.3). The fractional power emanating from the two output ports depends on the length L of the interaction region.

Let P_1 represent the power input into port 1 of the coupler, and let the power coming out of ports 2 and 3 (the output ports) be P_2 and P_3. The power P_4 coming out of port 4 is usually very small, and if the losses in the coupler (due to absorption, scattering, etc.) are small, the sum of the powers P_2 and P_3 is very nearly equal to P_1. The splitting of power between the second and third ports depends on the refractive index variation of the fibers, proximity of the two cores, the length L of the coupler, and the operating wavelength. If the two fibers are identical, the power exchange between the two fibers is complete, whereas if the two fibers are not identical, there is only incomplete power exchange (Fig. 12.4). In the case of couplers made of identical fibers, if power $P_1(0)$ is coupled into port 1, the power $P_T(L) = P_2$ exiting from port 2 and power $P_C(L) = P_3$ exiting from port 3 after an interaction over a length L are given by

$$\begin{aligned} P_2 &= P_1(0)\cos^2 \kappa L \\ P_3 &= P_1(0)\sin^2 \kappa L \end{aligned} \tag{12.1}$$

where κ, the coupling coefficient, is a function of the fiber parameters, the wavelength of operation, and the proximity of the fiber cores. Stronger coupling implies a larger value of κ, while weaker coupling implies smaller values of κ.

By an appropriate choice of κL, it is possible to achieve any fractional coupling between the fibers. Thus, it is possible to have an arbitrary fraction of power exiting from the coupled port (port 3). Thus, by choosing an appropriate length of the coupler, it is possible to achieve equal power outputs from the two ports (Fig. 12.5). Such a coupler is referred to as a 3-dB coupler since 3 dB corresponds to half (i.e., a coupling

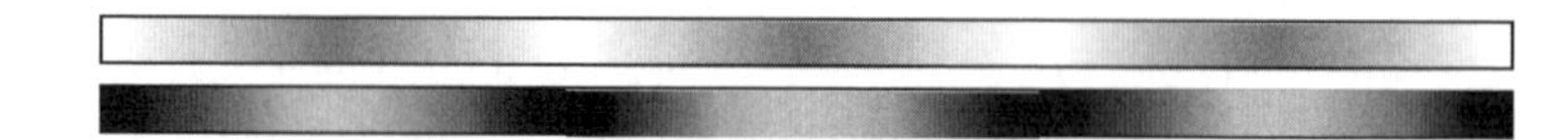

FIGURE 12.4 If the two fibers are nonidentical, the exchange of power is incomplete.

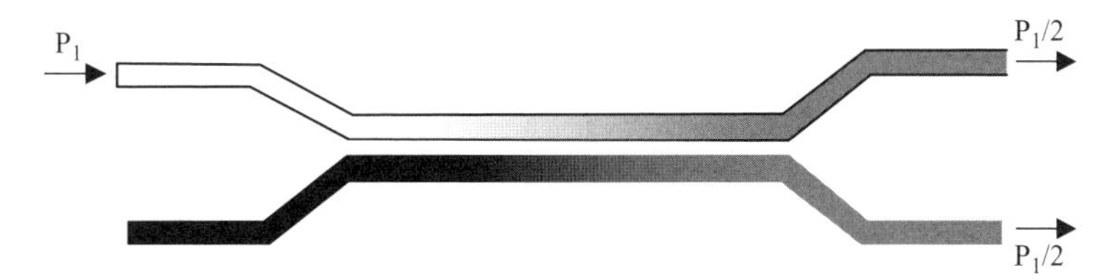

FIGURE 12.5 If the length of the coupler is chosen appropriately, the incident power can be split equally between the two output fibers. Such a coupler is called a 3-dB coupler.

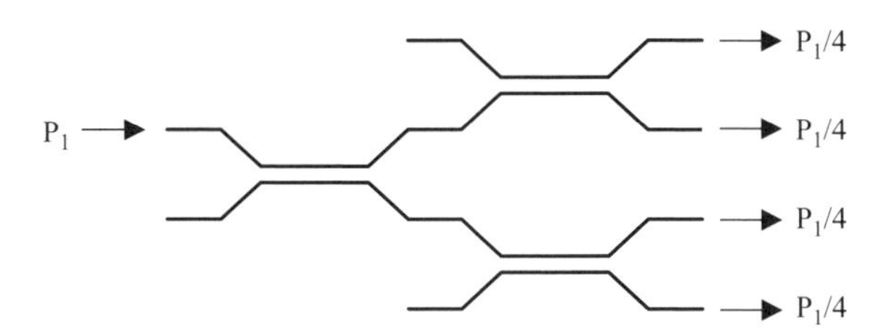

FIGURE 12.6 By concatenating a series of 3-dB couplers, it is possible to realize power splitters with one input port and multiple output ports carrying equal powers.

of 50%). Many 3-dB couplers can be connected one after the other to split the power further into multiple ports (Fig. 12.6).

Fiber optic couplers are usually specified by some important characteristics. These include:

- *The coupling ratio:* the ratio of the power in the coupled port to the total output power:

$$R(\text{dB}) = 10 \log \frac{P_3}{P_2 + P_3} \tag{12.2}$$

  Thus, for an input power of 1 mW, if the total output power is 0.9 mW and the coupled port carries a power of 0.1 mW, the coupling ratio is about 9.5 dB.

- *The excess loss:* the difference in power between the total output power and the input power, usually expressed in decibels:

$$L_{\text{ex}}(\text{dB}) = 10 \log \frac{P_1}{P_2 + P_3} \tag{12.3}$$

  Thus, if the excess loss is 1 dB, this implies that for an input power of 1 mW, the total output power (sum of powers from both the output ports) is about 0.79 mW; the remaining 0.21 mW is lost due to scattering, absorption, and the like.

- *The insertion loss:* the ratio of input power to coupled power:

$$I(\text{dB}) = 10 \log \frac{P_1}{P_3} \tag{12.4}$$

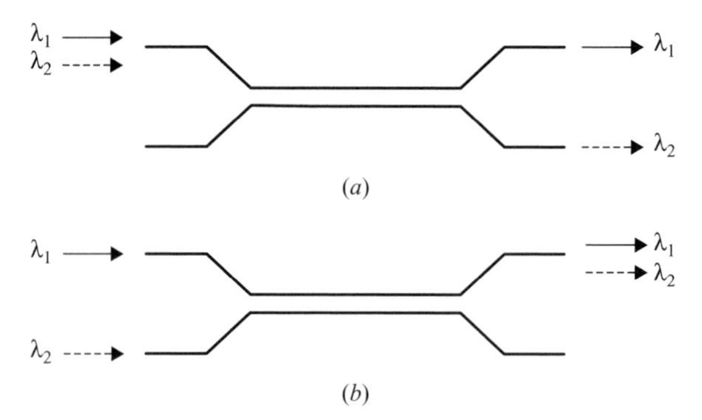

FIGURE 12.7 Wavelength-division demultiplexers and multiplexers can separate or combine two different wavelengths.

- *The directivity:* the power returning in the second input port, measured in decibels:

$$D(\text{dB}) = 10 \log \frac{P_4}{P_1} \tag{12.5}$$

If for an input power of 1 mW a power of 0.01 μW emerges from the second input port, the ratio of the two powers is 0.00001 and the directivity is −50 dB.

Tap couplers are couplers in which the fractional power appearing in the coupled port is a very small fraction of the light intput into the coupler. Thus, tap couplers tapping 1% or 5% of the light beam are used in many applications, such as in EDFAs.

Since the extent of field penetrating the cladding depends on wavelength, the coupling process is wavelength dependent (i.e., the coupling coefficient κ is wavelength dependent). Thus, it is possible to design couplers such that when power at two wavelengths is incident in the same port of a coupler, one of the wavelengths exits from one port while the other wavelength exits from the other port (Fig. 12.7). Such a coupler is used to multiplex (combine) or demultiplex (separate) two wavelengths. These are referred to as *wavelength-division-multiplexing* (WDM) *couplers* and find applications in EDFAs to combine the pump and signal power or to separate two wavelengths which are separated sufficiently, such as 1310 and 1550 nm.

Directional couplers have many interesting applications: for example, in power splitting, wavelength-division multiplexing and demultiplexing, polarization splitting, and fiber optic sensing.

Applications

Power Dividers As discussed earlier, one of the most important applications of a fiber directional coupler is as a power divider. In many applications, such as in local area networks or in fiber optic sensing, it is necessary to split or combine optical beams. Such a fiber optic directional coupler forms an ideal component since it is compact and possesses low loss. One of the important issues in such couplers is the wavelength dependence of the coupler. Wavelength-flattened couplers are characterized by coupling ratios, which remain almost constant for a given band of wavelengths, such as the C-band (1530 to 1565 nm). If power is to be divided into many ports, 3-dB couplers can be concatenated to split from one input port into multiple ports, as shown in Fig. 12.6.

Wavelength-Division Multiplexers/Demultiplexers Another very important application of such couplers is in wavelength-division multiplexing and demultiplexing. As discussed earlier, fiber directional couplers are in general wavelength sensitive, and we may have a coupler for which if light beams at two different wavelengths (say, one at 980 nm and the other at 1550 nm) are launched simultaneously in one of the ports, one of the wavelengths will exit port 2 and the other wavelength will exit from the other port, as shown in Fig. 12.7*a*. If we use the same coupler in the reverse direction, the two wavelengths incident on the coupler appear at the same port, as shown in Fig. 12.7*b*; such a coupler is used in an erbium-doped fiber amplifier as shown in Fig. 9.6.

Coupler Fabrication

Fiber couplers are fabricated using a variety of techniques; however, the method used most extensively is fusion. Fused fiber couplers are fabricated by first slightly twisting two single-mode fibers (after removing their protective coating) and then heating and pulling them so that the fibers fuse laterally with one another and are also tapered (Fig. 12.8). Heating can be accomplished by using an oxybutane flame or miniature electrical heating element. The coupling ratio is monitored online as the fibers are fused and drawn, and the process of fusion and tapering is stopped as soon as desired. Figure 12.9 shows the cross section of a fused fiber coupler composed of single-mode fibers. One can see the two cores that lie close to each other.

12.5 ISOLATORS

In many applications it is desired to have an optical fiber component through which light should be able to propagate along one direction but not in the reverse direction.

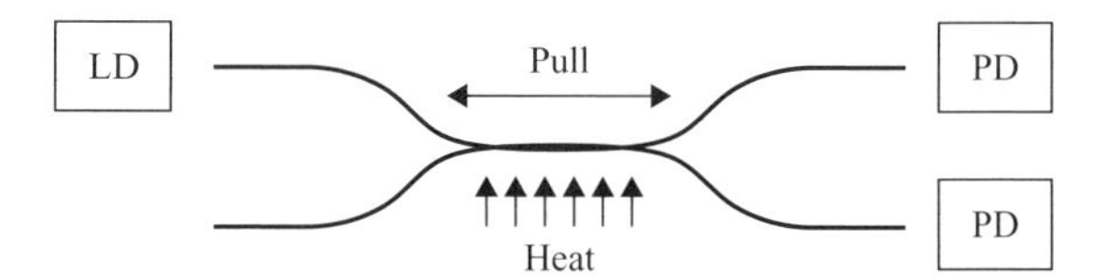

FIGURE 12.8 Fused fiber coupler fabrication involves bringing two fibers close together and then heating and pulling them. Online monitoring of the outputs from both output ports is used to control the splitting ratio. LD, laser diode; PD, photodiode.

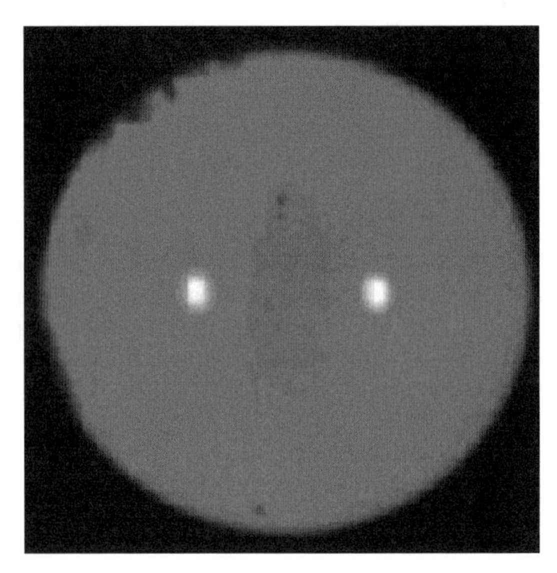

FIGURE 12.9 Cross section of a fused coupler in the coupling region. (After Pone et al., 2004.)

Such devices are called *isolators*, as they isolate the input portion and from the output portion from any reflections that may be occurring in the path of the output arm. Such isolators are a very important component in any EDFA (Chapter 9).

The basic principle used in the realization of such a component is the Faraday effect, discussed in Chapter 14. In such an effect, when linearly polarized light propagates through an optical element that is subjected to a magnetic field applied along the direction of propagation of the light beam, the plane of polarization of the light beam rotates (Fig. 12.10). The angle of rotation depends on the material and on the magnitude of the magnetic field and the length of the medium. The Faraday effect is referred to as a *nonreciprocal effect* since if we now send a beam in the reverse direction, then since the light wave is now propagating opposite to the direction of the applied magnetic field, the rotation of the plane of polarization takes place along the same direction and the output beam from the input side is now not parallel to the input polarization (Fig. 12.10).

Now if the applied magnetic field is such as to cause a rotation of the plane of polarization by 45°, any reflected wave as it traverses through the medium will exit with its state of polarization perpendicular to the input polarization state. If we place a polarizer in the input path, the reflected light, having its polarization state normal to the polarizer, will not be able to pass through, and thus any reflection from the output end will be blocked. Figure 12.11 shows an optical isolator that allows light to propagate from left to right but stops any beam traveling from right to left. Such isolators are usually provided with fiber optic input and output ports (pigtails) and

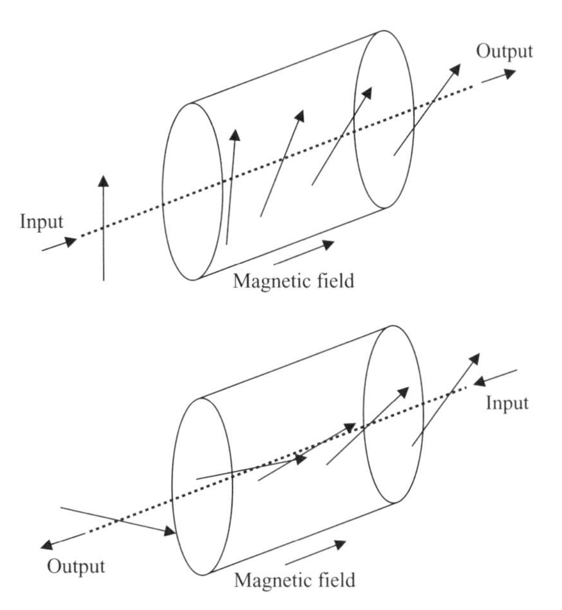

FIGURE 12.10 In the Faraday effect, the plane of polarization of a light beam rotates in the presence of a magnetic field applied along the direction of propagation. If the rotation is 45°, then when the light travels back along the same medium, the plane of polarization of the outcoming beam is perpendicular to that of the input.

are also very compact. They are usually characterized by a loss of 0.2 to 2 dB in the forward direction, while the loss in the reverse direction could be 20 to 40 dB.

The principle used in making an isolator can also be used to build a *circulator*, in which light entering one port (say, port 1) leaves the device via another port (say, port 2), and light that enters from port 2 exits from another port (say, port 3) (Fig. 12.12). Circulators are very useful devices in many applications, such as in dispersion compensation using fiber Bragg gratings (Chapter 11).

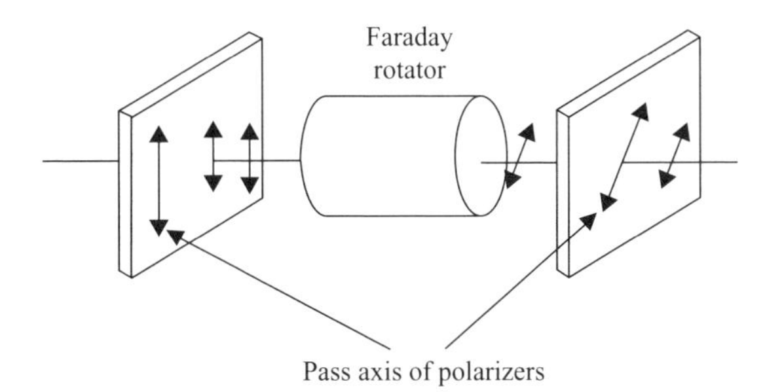

FIGURE 12.11 Optical isolator using the Faraday effect. Light can propagate from left to right but not in the reverse direction through the device.

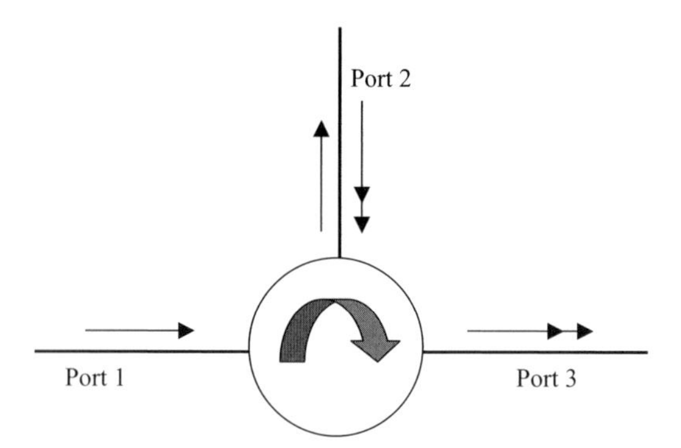

FIGURE 12.12 Optical circulator.

12.6 ARRAYED WAVEGUIDE GRATINGS

In WDM systems a number of wavelength channels carry independent signals through the optical fiber. Thus, there is a need to multiplex and demultiplex a number of wavelengths, and arrrayed waveguide gratings are devices that can perform this operation very efficiently with low losses. AWGs capable of multiplexing up to 1080 channels with 25-GHz spacing have also been demonstrated. The AWG is a planar device (usually fabricated on a silicon substrate) and consists of a series of input waveguides connected to a planar waveguide that is coupled to an arrayed waveguide structure as shown in Fig. 12.13. The arrayed waveguides are then connected to a series of output waveguides via another planar waveguide. The adjacent waveguides of the arrayed waveguide system differ in length by fixed amounts. Light at different wavelengths input into one of the input waveguides gets diffracted in the planar waveguide and gets coupled to the waveguides of the arrayed waveguide structure. Since the individual waveguides of the arrayed waveguide structure are of different lengths when the light waves arrive at the second planar waveguide, they interfere such that different wavelengths get focused at different points and subsequently get

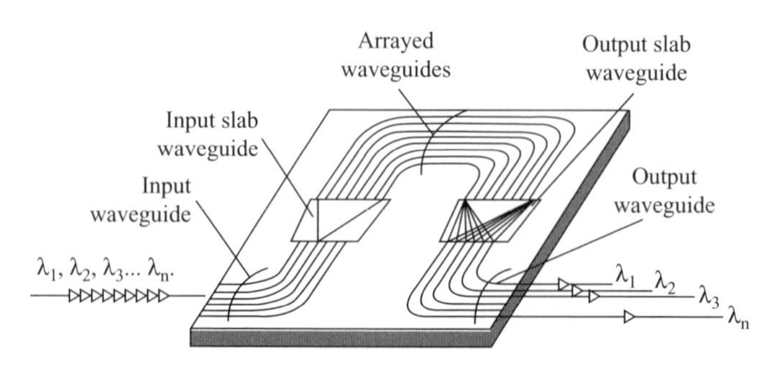

FIGURE 12.13 Arrayed waveguide grating device.

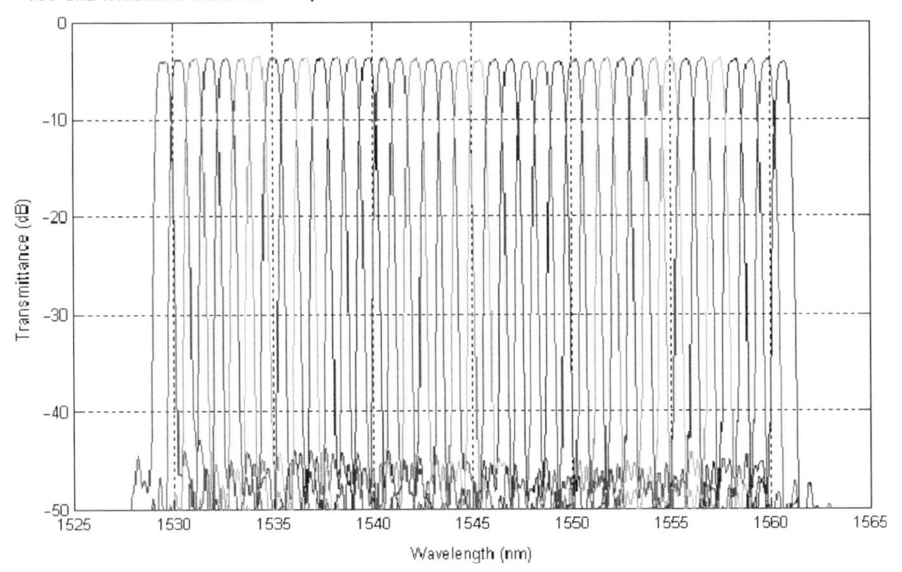

FIGURE 12.14 Demultiplexed output from an AWG. (Adapted from JDS Uniphase; http://www.jdsu.com.)

coupled to different output waveguides. Thus, the various wavelengths that were coupled into a single input waveguide get demultiplexed and exit from different outputs. If this device operates in reverse, it will act as a multiplexer.

Figure 12.14 shows the demultiplexed output from different waveguides of an AWG; the channels are spaced by as little as 50 GHz (corresponding to a wavelength spacing of 0.4 nm). The insertion loss of the device is typically less than 5 dB, the wavelength accuracy is typically better than 0.02 nm, and the bandwidth is about 0.2 nm.

12.7 EXTERNAL MODULATORS

We mentioned earlier that laser diodes can be modulated directly by varying the current through them. Thus, a bit stream in the electrical domain can be converted directly into a bit stream in the optical domain by direct modulation. When this is carried out, it is found that the wavelength of the laser gets chirped within the pulse, due to the variation of electron concentration within the laser. Since the spectral bandwidth of a chirped laser is larger than that of an unchirped laser, and dispersion through a fiber depends on the bandwidth, the dispersion suffered by a pulse formed by a chirped laser is greater than that due to an unchirped laser. This increase is not very significant at bit rates of 2.5 Gb/s, but for higher bit rates, such as 10 and 40 Gb/s, it can affect the repeater spacing considerably. In view of this, for higher bit rates an external modulation scheme is used rather than direct modulation. In this scheme the laser operates continuously and passes through an external modulator that converts the continuous wave into pulses according to the data. Many different methods are used to achieve this external modulation. Here we describe the most common technique using an electrooptic modulator.

There are crystals such as lithium niobate whose refractive index can be changed by an applied electric field. For an applied electric field E, the change in refractive index Δn of the crystal is given by

$$\Delta n = \frac{n^3 r E}{2} \tag{12.6}$$

where r, the effective *electrooptic coefficient*, depends on the material and on the polarization states of the light and the direction of applied electric field within the crystal, and n represents the refractive index of the material in the absence of an external electric field.

As an example we consider lithium niobate, for which if the electric field is applied along a specific direction (referred to as the *optic axis*) and if the incident light is polarized along the direction of the optic axis, $r = 30 \times 10^{-12}$ m/V and $n \simeq 2.2$. Hence, if a potential difference of 100 V is applied across a crystal of thickness 1 mm, the electric field generated is 10^5 V/m. For this applied voltage the change in refractive index is approximately 1.6×10^{-5}, which is indeed a small shift. If the light beam propagates through such a crystal, this change in refractive index would lead to a change in the phase of the light beam passing through the crystal. If the wavelength of light is 1.5 μm and the crystal length is 2 cm, the change in phase due to application of this electric field would be about 0.4π, which is not a small change if we remember that in interference the intensity of the interference pattern can change from maximum to minimum for a change of phase of π.

An electrooptic modulator utilizes this change in refractive index due to an applied electric field to modulate a light beam. Hence, when the refractive index is modulated using the electric field, the phase of the output light beam also gets modulated, and if the output is made to interfere with a second beam, which has not undergone modulation, the interference between the two waves will lead to modulation of the resulting intensity. When the waves are in phase (the crest of one wave overlapping with the crest of the other wave) they will add constructively, and when they are out of phase (the crest of one wave overlapping with the trough of the other wave), they will add destructively. Thus, if the electric field is switched on and off in accordance with the signal bits, the output light wave from the device will replicate the bit sequence. This is the basic principle of an electrooptic amplitude modulator.

The modulation described above is generally achieved using a Mach–Zehnder interferometer, shown in Fig. 12.15 (see also Chapter 14). An input waveguide splits into two waveguides which recombine after a certain length. An electric field is applied to one of the waveguides using a pair of electrodes. When no voltage is applied, the output from both arms come in phase and results in output from the device. When an appropriate voltage is applied to the electrode to make the

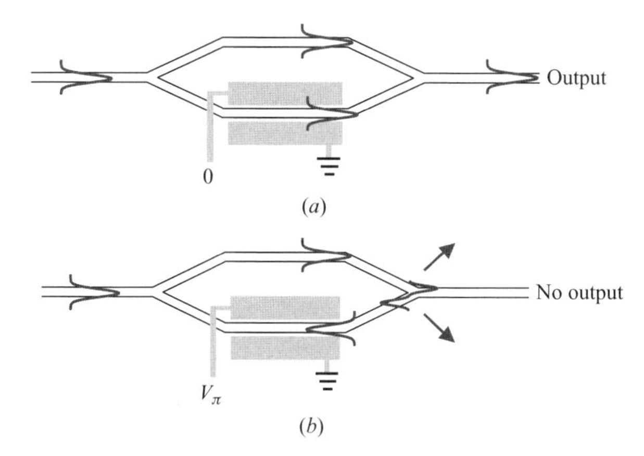

FIGURE 12.15 Mach–Zehnder interferometric modulator. If no electric field is applied on the electrodes, the input light exits from the output. When a specific voltage is applied, the output becomes zero.

two waves interfere destructively, there is no output from the device. Figure 12.16 shows a typical variation of output intensity with applied voltage showing sinusoidal dependence. Figure 12.17 shows a typical intensity modulator connected using fiber optic pigtails. Such modulators can be used to modulate at very high speeds of 10 and 40 Gb/s. They typically require voltages of less than 5 V for operation and are used in high-speed fiber optic systems.

Figure 12.18 shows a thermooptic modulator based on a Mach–Zehnder interferometer based on silicon technology. The waveguides used to guide light are made of silica and are built on a silicon substrate. By placing heaters on the waveguides,

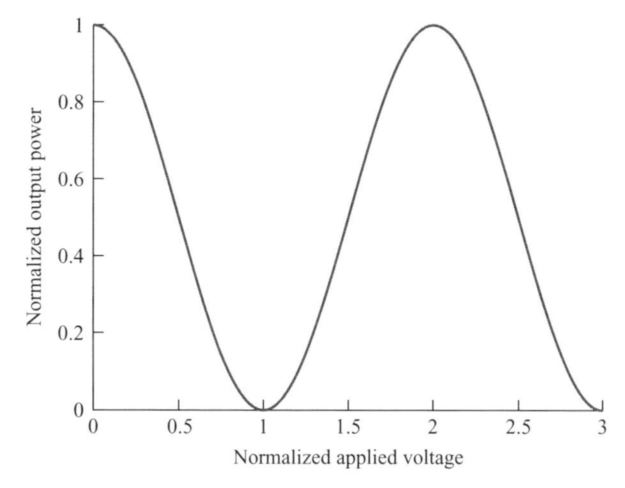

FIGURE 12.16 Variation of normalized output intensity with applied voltage for a Mach–Zehnder interferometric modulator.

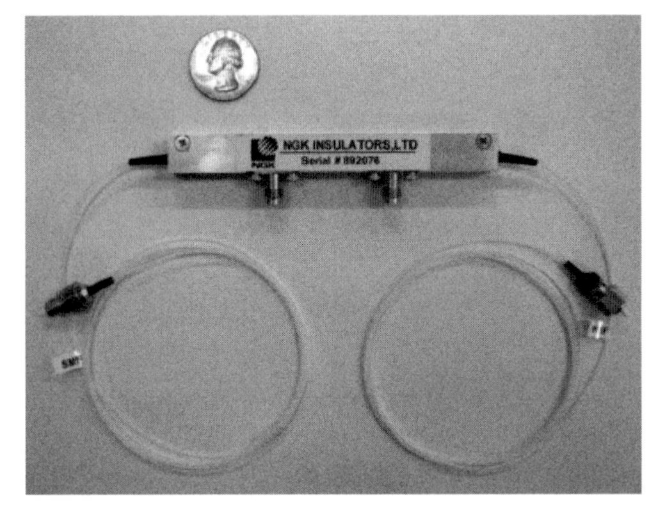

FIGURE 12.17 Typical electrooptic modulator used in fiber optic communication systems. (Adapted from NGK Insulators Ltd.; http://www.ngk.co.jp.)

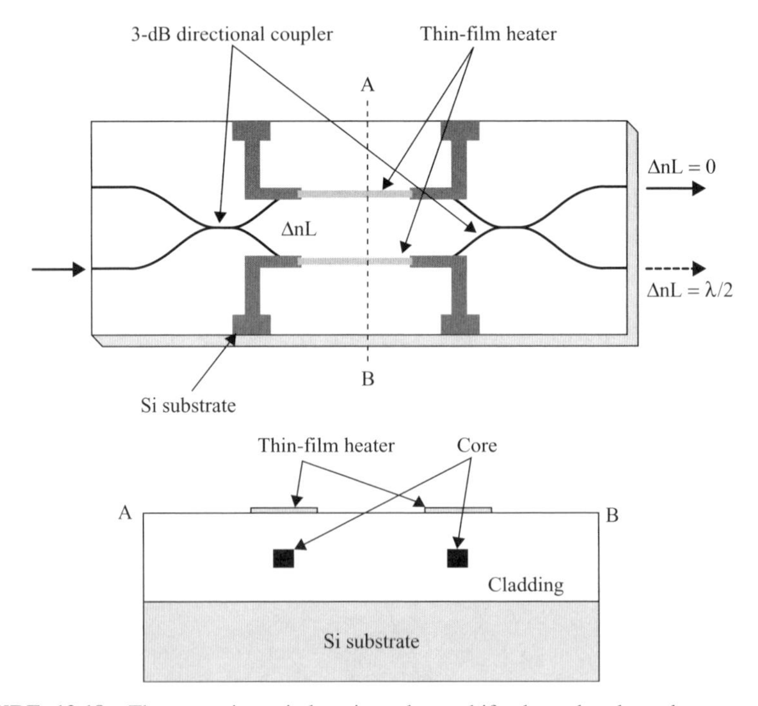

FIGURE 12.18 Thermooptic switch using phase shifts brought about by temperature changes in the waveguide.

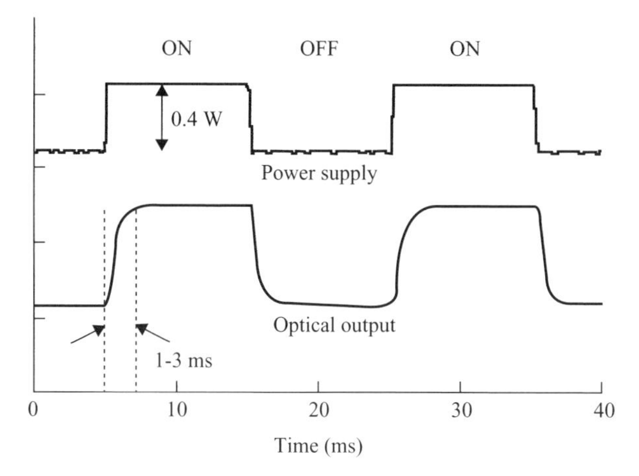

FIGURE 12.19 Typical response of a thermooptic switch.

the phase of light propagating in the waveguides can be changed by temperature change brought about by the heating elements. This change then causes a change in the output intensity. Of course, unlike electrooptic modulators, such devices are slow since thermal effects are much slower then electrooptic effects. Figure 12.19 shows a temporal response of the device, showing a switching time of a few milliseconds; the corresponding power requirement is less than half a watt.

Figure 12.20 shows the eye pattern of the output from a modulator operating at 40 Gb/s. The drive voltage required for the device is only 4.1 V, and the insertion loss of the device is 5.4 dB. Thus, such electrooptic modulators can help in very high speed modulation of laser diodes and are being used in high-speed optical fiber communication systems.

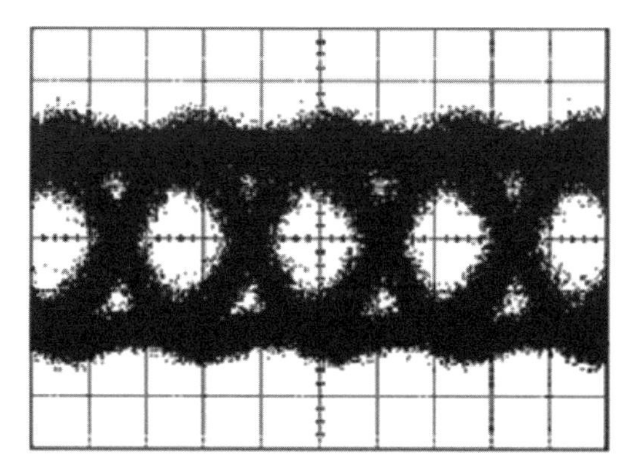

FIGURE 12.20 Eye diagram of a 40-Gb/s bit pattern from an external modulator.

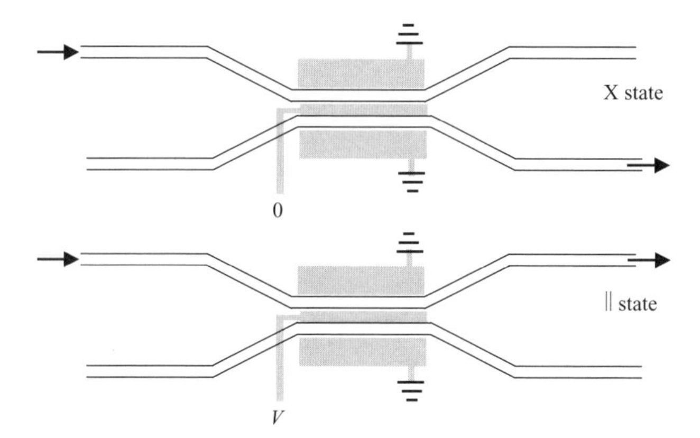

FIGURE 12.21 Optical switch based on a directional coupler. Such switches are fabricated in substrates such as lithium niobate or in semiconductors.

Optical Switches

Using the electrooptic effect discussed above, it is possible to achieve optical switching. Figure 12.21 shows a directional coupler switch. It comprises a directional coupler that consists of two identical single-mode waveguides fabricated on a substrate such as lithium niobate on which are placed metal electrodes so as to apply an electric field. When no electric field is applied, the directional coupler length is adjusted so that all the light emerges from the coupled port of the coupler. This is referred to as the *cross state* (the × *state*). When an electric field is applied, the waveguides become nonidentical since the direction of the electric field is opposite in the two waveguides. This results in an incomplete power transfer between the waveguides, and with an appropriate voltage, the output light can be made to emerge from the upper waveguide, referred to as the *parallel state* (|| state). Figure 12.21 shows the parallel and cross states of a directional coupler switch. It is possible to concatenate such switches to form switch matrices for switching between multiple ports at the input to multiple ports at the output.

Wavelength Converters

Wavelength converters are very useful components in a wavelength-routed optical fiber network. In such networks at some nodes it is required to convert information bits at one wavelength to identical bits at another wavelength. Wavelength converters are expected to perform this function. Wavelength converters should be able to convert the wavelength at high bit rates, such as 10 Gb/s and higher. The converted wavelength should have a good extinction ratio (ratio between 1's and 0's). There are various types of wavelength converters. The simplest is first to convert the incoming optical signals into electrical signals and then use the electrical signal to drive another laser diode at a different wavelength. Of course, this would need very high speed electronic circuitry for high-speed wavelength conversion, and conversions of multiple channels into another set of multiple-wavelength channels would require multiple units, which would make this solution very expensive. To achieve this, all-optical wavelength converters use cross-gain modulation in semiconductor optical amplifiers or four-wave mixing.

In the case of semiconductor optical amplifiers, the input signal bits and a continuous probe beam are simultaneously incident on the SOA. The probe beam modulates the gain of the SOA, due to its fast gain dynamics, and this gain modulation leads, in turn, to a variation in the output power in the continuous-wave probe signal. Using this scheme, conversion at rates of 40 Gb/s have been demonstrated.

12.8 THIN-FILM DEVICES

One very common thin-film device is the wavelength multiplexer or demultiplexer. This is based on interference occurring due to multiple reflections. Figure 12.22 shows a device called a Fabry–Perot interferometer, which consists of a pair of highly reflecting mirrors placed parallel to each other. When light is incident on this device, each mirror reflects a major fraction of this light, and light bounces back and forth between the pair of mirrors. Wavelengths satisfying the following condition get transmitted completely (assuming no other loss mechanism) through the device:

$$\lambda_l = \frac{2nd}{l} \qquad l = 1, 2, 3, \ldots \tag{12.7}$$

Here n represents the refractive index of the medium between the two mirrors and d is the spacing between the mirrors. If the mirror reflectivities are high, the transmission maxima are very sharp, and any deviation from the wavelengths specified by Eq. (12.7) leads to almost no transmission, and all those wavelengths are almost completely reflected back.

Such an effect is used to build wavelength filters. Figure 12.23 shows a wavelength filter with multiple-layer coatings. The series of multiple layers of alternate dielectric media of low and high refractive indices act as high-reflectivity mirrors, and this leads to a wavelength filtering effect. Fiber Bragg gratings (Chapter 11) use the same effect to reflect strongly at desired wavelengths. Figure 12.24 shows how such a

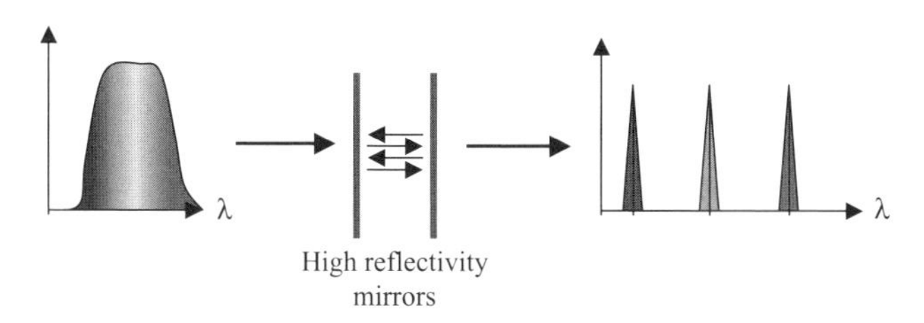

FIGURE 12.22 When light falls on a pair of parallel high-reflectivity mirrors, only certain wavelengths satisfying Eq. (12.7) get transmitted, other wavelengths get reflected almost completely.

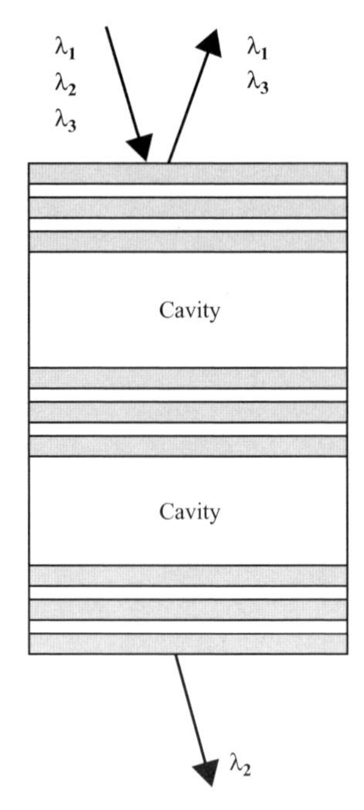

FIGURE 12.23 Realization of high-reflectivity mirrors using multiple coatings of dielectrics of different refractive indices.

device can be configured to make wavelength multiplexers and demultiplexers. Such devices, which are available commercially for low channel counts have insertion losses between 1 and 5 dB, an adjacent channel isolation better than 25 dB, and a directivity better than 45 dB. They are also made to be thermally very stable by proper packaging.

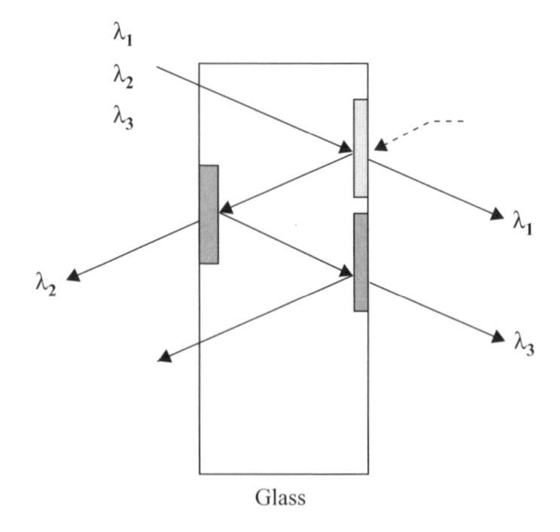

FIGURE 12.24 Using the interference effect, it is possible to realize wavelength multiplexers and demultiplexers.

CHAPTER 13

Nonlinear Effects in Optical Fibers

13.1 INTRODUCTION

Who has not enjoyed standing on a beach caressed by the breaking waves? Surface waves usually originate from wind blowing over the water surface. As the waves approach the sloping beach, their amplitude increases and they slow down. Unlike the waves that we discussed in Chapter 2, the speed of these waves depends on the amplitude. Thus, the larger-amplitude portion of the wave travels faster than the smaller-amplitude portion, and eventually the wave breaks (Fig. 13.1). The dependence of the wave speed on the amplitude of the wave is a characteristic nonlinear phenomenon.

In the case of light waves, when we are dealing with light from ordinary sources with small powers, the propagation of light waves in any medium is linear. This means that the propagation effects of a light wave are independent of the intensity of the light wave. The characteristics of the emerging light beam such as frequency, phase, and wave shape, remain unchanged as we change its intensity. Apart from this, when two light beams propagating through a medium intersect, each beam emerges without any modification brought about by the other beam. Thus, referring to Fig. 13.2, we see that there would be no change in the output beam 1 whether or not beam 2 is present, and conversely. Thus, one light beam does not modify any of the properties of another light beam, even if they cross each other. This happens whenever the intensities of the two light beams are small, and this is referred to as *linear optics*.

When the light intensity becomes large, the electric field associated with the light beam can modify the property of the medium to such an extent that it can then affect its own propagation as well as that of other beams crossing it. For example, a red beam entering an appropriate crystal can get converted partly to a blue output beam under certain conditions (Fig. 13.3). This happens due to a nonlinear effect (called *second harmonic generation*) in which a light beam of frequency f creates a beam having double its frequency, $2f$ (half the wavelength). Similarly, the refractive index of the medium can get changed due to the large intensity of the beam; this change of refractive index would in turn change the phase with which a light wave emerges from a medium. Such effects, referred to as *nonlinear optical effects*, have

Fiber Optic Essentials, By K. Thyagarajan and Ajoy Ghatak

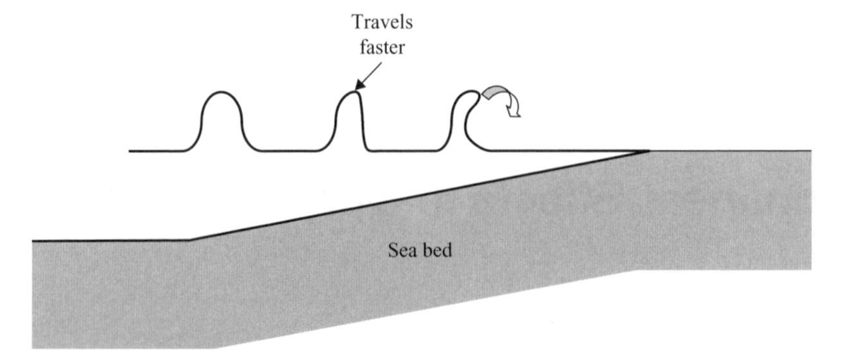

FIGURE 13.1 Wave breaking on a beach due to the dependence of the wave speed on the wave amplitude.

become very important since discovery of the laser. The nonlinear optical effects are almost instantaneous: that is, any change in intensity of the beam leads almost instantaneously to a change in the medium property, and the medium relaxes to its original state within times less than a few femtoseconds (1 femtosecond = 10^{-15} s).

In optical fiber communication, nonlinear optical effects start to play an important role whenever the intensity levels are high, as when the number of wavelength channels carrying signals increases or optical amplifiers are used to amplify the optical signals so that the repeater distance can be increased. Due to the nonlinear effects taking place within the optical fiber, the signal pulses carrying information can get modified due to the presence of other channels, which in turn can lead to increased errors in detection. Since the nonlinear effects depend on the intensity of the light wave, for a given light power as we reduce the cross-sectional area of the beam, nonlinear effects become stronger and stronger. In the case of optical fibers, the optical beam is confined to propagate within the core, and if the area of cross section of the beam is small, stronger nonlinear effects can be observed. This has led to using nonlinear optical effects in optical fibers for all-optical processing of optical signals (i.e., using optical effects to modify the properties of light beams). All optical signal processing is being pursued as an efficient way for processing of high-speed optical signals, such as for switching and wavelength conversion.

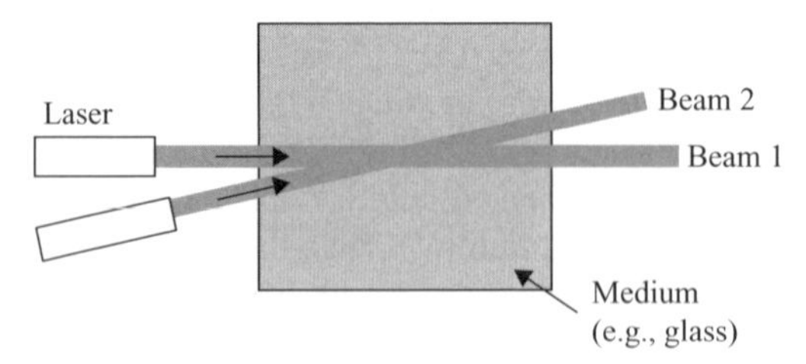

FIGURE 13.2 For a linear medium when two light beams cross, they travel independently and have no effect on the propagation of the other beam.

FIGURE 13.3 Second harmonic generation, in which a red laser beam enters a crystal from the right and comes out from the left as a blue beam having double the frequency of the incident beam. (Courtesy of Dr. R. W. Terhune.)

In this chapter we discuss some of the significant nonlinear optical effects that are important primarily in optical fiber communication systems. With the increase in number of channels and repeater spacing, these nonlinear effects would finally limit the transmission capacity of the link. Detailed discussions of nonlinear effects in optical fibers can be found in Agarwal (1989).

13.2 SELF-PHASE MODULATION

When we refer to the refractive index of a medium, we never specify the intensity levels since under normal circumstances the refractive index of a medium is independent of the light intensity. On the other hand, at high optical intensities, the refractive index of a medium is no longer a constant but depends on the intensity of the light beam. If the refractive index of a medium for low intensities is n_0, at high intensities it can be written as

$$n = n_0 + n_2 I \tag{13.1}$$

where n_2 is a constant that depends on the medium and I represents the intensity of the beam. For example, for silica, $n_2 \approx 3.2 \times 10^{-20}$ m^2/W, and thus the refractive index of silica increases with increased light intensity; for nominal intensities the increase is, of course, extremely small. Thus, if we consider a single-mode optical fiber with a mode area of 80 μm^2 and couple light of power 100 mW, the intensity within the fiber would be about 1.25×10^9 W/m^2 and the corresponding increase in refractive index due to the presence of the light beam is a tiny value of about 4×10^{-11} (i.e., 0.00000000004), so the refractive index would increase from 1.44 to 1.44000000004. Although this refractive index change is very small, when the light beam propagates in an optical fiber over long distances (a few hundred to a few thousand kilometers), the accumulated effects due to this increase can be significant. Since the phase of the light beam depends on the refractive index of the medium, and it is the beam itself that is changing the refractive index, which in turn changes its own phase, this effect is referred to as *self-phase modulation*.

In a single-mode optical fiber, since the light beam propagates as a mode, we can represent the intensity as a ratio of the power carried by the beam to the mode area (A_{eff}). Thus, in the case of an optical fiber, Eq. (13.1) is written as

$$\tilde{n}_{\text{eff}} = n_{\text{eff}} + n_2 \frac{P}{A_{\text{eff}}} \tag{13.2}$$

where n_{eff} is the effective index of the mode for low powers (when nonlinearity is negligible) and $\tilde{n}_{\text{eff}}$ is the effective index in the presence of nonlinearity. The effective mode area is given approximately by

$$A_{\text{eff}} = \frac{\pi\,(\text{MFD})^2}{4} = \pi w_0^2 \tag{13.3}$$

where MFD ($= 2w_0$) is the mode field diameter of the fundamental mode of the fiber and w_0 is the Gaussian spot size (see Chapter 4). Thus, fibers having a smaller mode area would exhibit larger nonlinear effects. Table 13.1 gives the mode areas of some common fibers.

It follows from Eq. (13.2) that in the presence of nonlinearity, the effective index changes by $n_2 P/A_{\text{eff}}$. This change in effective index will bring about a change in phase given by

$$\Delta\phi = k_0 n_2 \frac{P}{A_{\text{eff}}} = \gamma P \tag{13.4}$$

where

$$\gamma = \frac{k_0 n_2}{A_{\text{eff}}} \tag{13.5}$$

TABLE 13.1 Mode Area and Nonlinear Coefficient for Some Common Fiber Types

Fiber Type	Mode Area (μm^2)	Nonlinear Coefficient γ at 1550 nm ($W^{-1}\,m^{-1}$)
G.652	85	1.5×10^{-3}
G.653	46	2.8×10^{-3}
G.655	52 ($D > 0$)	
	56 ($D < 0$)	2.5×10^{-3}
NZ-DSF	73	1.8×10^{-3}
DCF	23	5.6×10^{-3}
PCF	3	43×10^{-3}

is the nonlinear coefficient and is a parameter that depends on the material through n_2, the wavelength through k_0 ($= 2\pi/\lambda_0$, λ_0 being the wavelength in free space) and the effective area through A_{eff}. Larger values of γ imply larger nonlinear effects. Table 13.1 gives values of A_{eff} and γ for some typical fibers.

Since nonlinearity-induced phase change depends on optical power, and since the optical power in the pulse decreases due to attenuation in the fiber, the effect of nonlinearity would decrease continuously as light propagates through the fiber. When light enters the fiber, it has maximum power, and this results in a high nonlinear phase shift. But as light propagates, its power decreases and the corresponding nonlinear effect would decrease. This effect is taken care of by defining an effective fiber length L_{eff}:

$$L_{\text{eff}} = \frac{1 - e^{-\alpha L}}{\alpha} \tag{13.6}$$

where α is the attenuation coefficient of the fiber and L represents the actual length of the fiber. If the fiber length $L \gg 1/\alpha$, the effective length is approximately $1/\alpha$. Similarly, if the length $L \ll 1/\alpha$, the effective length is approximately equal to the actual length L. For single-mode fibers operating at 1550 nm, $\alpha \sim 0.25$ dB/km (which in units of km^{-1} is approximately $5.76 \times 10^{-2}\ \text{km}^{-1}$), and thus $L_{\text{eff}} \sim L$ for $L \ll 17$ km and $L_{\text{eff}} \sim 17$ km for $L \gg 17$ km.

To find out about the effect of self-phase modulation on the propagation of an optical pulse, let us consider the propagation of an optical pulse through an optical fiber. Figure 13.4 shows the electric field variation of an optical pulse; you can imagine this to be like a snap shot of the pulse at any given time. The oscillatory variation is due to the high frequency of the light signal, while the envelope of the pulse (which defines the pulse shape) has a much slower variation with time. Now, in the leading and trailing edges of the pulse, the intensity is smaller than at the center

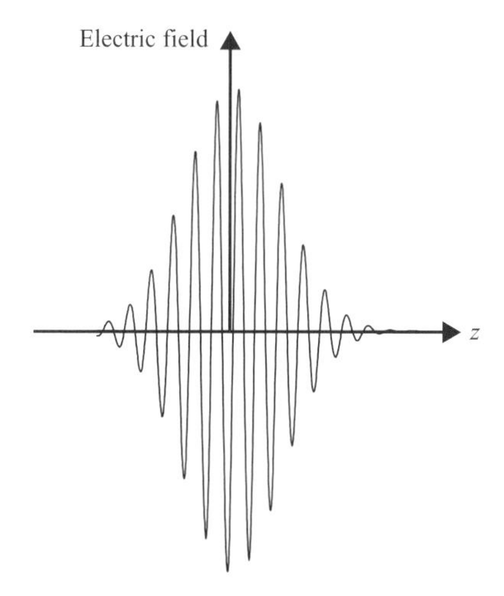

FIGURE 13.4 Optical pulse; the oscillatory portion is due to the high frequency of the pulse and the envelope is the pulse shape.

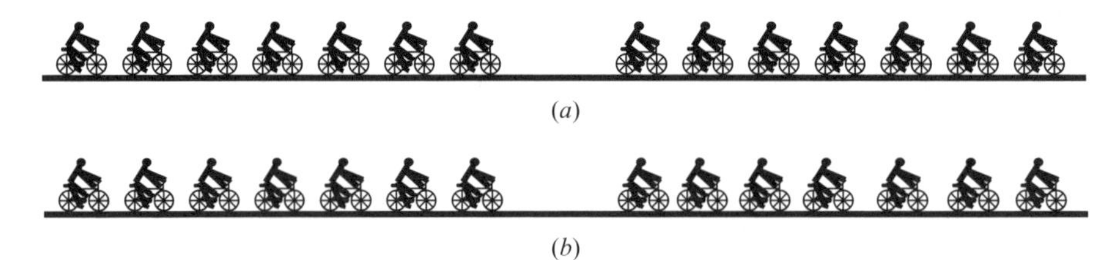

(a)

(b)

FIGURE 13.5 (*a*) All cyclists travel at the same speed, and their separation remains the same as they propagate. (*b*) If the cyclists close to the center start to travel slightly slowly, this would result in crowding at the back and greater separation at the front.

of the pulse. Hence, due to the dependence of refractive index on intensity, the center of the pulse would lead to a greater increase in the refractive index of the core of the fiber than will the leading and trailing edges. Since the response of the medium to changes in intensity are almost instantaneous (a short response time compared to the time period of the optical wave, which is in femtoseconds) as the intensity of the pulse changes within the pulse, the refractive index change follows the change in intensity almost instantaneously. As the speed of propagation of light depends on the refractive index, this would result in a slight slowing down of the center of the pulse vis-à-vis the leading and trailing edges.

To understand the implication of this, we consider an example in which seven cyclists are traveling along a road at the same speed and spaced equally (Fig. 13.5). In this case the group of cyclists will retain their positions as they travel. Now, if the cyclists closer to the center of the group start to travel a bit slower than the rest, there would be a crowding of the cyclists toward the back end and greater separation toward the front end.

In the case of light pulse propagation through an optical fiber, we can imagine a similar situation when the higher-intensity portion of the pulse (around the center of the pulse) travels more slowly than the ends. There would then be a crowding of the waves toward the back end and greater separation toward the front end. Since the period of oscillation decides the frequency of a wave, this implies that the front end of the pulse would have a lower frequency and the back end of the pulse would have a higher frequency (Fig. 13.6). In such a pulse, called a *chirped pulse*, the chirping is caused by nonlinear effects. Chirping without a change in pulse shape leads to an increase in the frequency content of the pulse, that is, to a broadening of the spectrum of the pulse.

The instantaneous frequency of the pulse is given by

$$\omega(t) = \omega_0 - \gamma L_{\text{eff}} \frac{dP_0}{dt} \qquad (13.7)$$

where P_0 is the time-dependent power variation of the pulse and ω_0 is the center frequency of the incident pulse. The leading edge of a pulse corresponds to a positive value of dP_0/dt, and hence the frequency is lower than ω_0 while the trailing edge

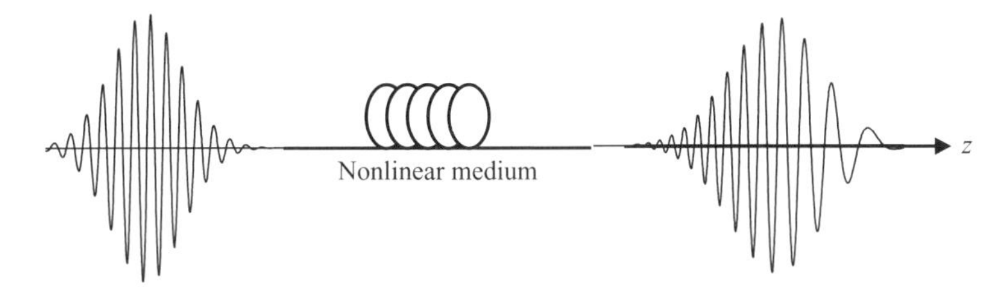

FIGURE 13.6 When an optical pulse travels through an optical fiber in the presence of a nonlinear effect, the frequency of the pulse varies with position within the pulse, leading to a chirped pulse.

corresponds to a negative value of dP_0/dt, and hence the frequency is higher than ω_0 (Fig. 13.6). Figure 13.7 shows simulation results of pulse propagation through an optical fiber in the presence of nonlinearity only (neglecting dispersion), showing clearly that the pulse shape remains the same while the frequency spectrum broadens.

We saw in Chapter 7 that the dispersion of a pulse depends on the spectral bandwidth of the pulse (i.e., on the frequency content of the pulse). Hence, in the presence of such a nonlinear effect, the dispersive behavior of the pulse would get changed. It so happens that when we operate a standard single-mode fiber (with zero dispersion close to 1310 nm) at 1550 nm (i.e., a fiber operating with positive dispersion), the presence of nonlinear effects indeed results in a reduction of the effective dispersion. This implies that the dispersion caused in the pulse would decrease as the input power is increased. Hence, the quantum of dispersion compensation required in the presence of nonlinearity is in fact less than what is predicted using linear effects only. This fact has to be taken into account when designing a fiber optic system. For operation below the zero-dispersion wavelength, wherein the dispersion is negative, the nonlinear effect indeed leads to increased dispersion.

In the special class of fibers called *photonic crystal fibers* and *holey fibers*, the cross-sectional area of the light propagating in the core can be as small as 2.5 μm^2, compared to standard single-mode fibers, in which the cross-sectional area is about 75 μm^2. Figure 13.8 shows the cross section of a holey fiber and an expanded photograph of the central section. As can be seen, the fiber consists of a regular pattern of holes surrounding a very small core and lying all along the length of the fiber, with hole diameters of less than 1 μm. The effect of the holes is to tightly confine the light beam into the very small cross section of the small central core, resulting in a very small area of cross section of the propagating light beam. Thus, for a given power within the fiber, the intensity levels (which is power carried per unit cross-sectional area) would be much higher in the case of such fibers. The intensity of the light within the fiber can become very significant, even with small coupled powers, leading to enhanced nonlinear effects. Figure 13.9 shows how light at a single wavelength (of about 850 nm) coupled into a holey fiber can lead to the generation of an entire spectrum of light, spanning from about 400 to 1600 nm. Such an effect, referred to as *supercontinuum generation*, finds applications in spectroscopy, ultrafast pulse generation, and other fields.

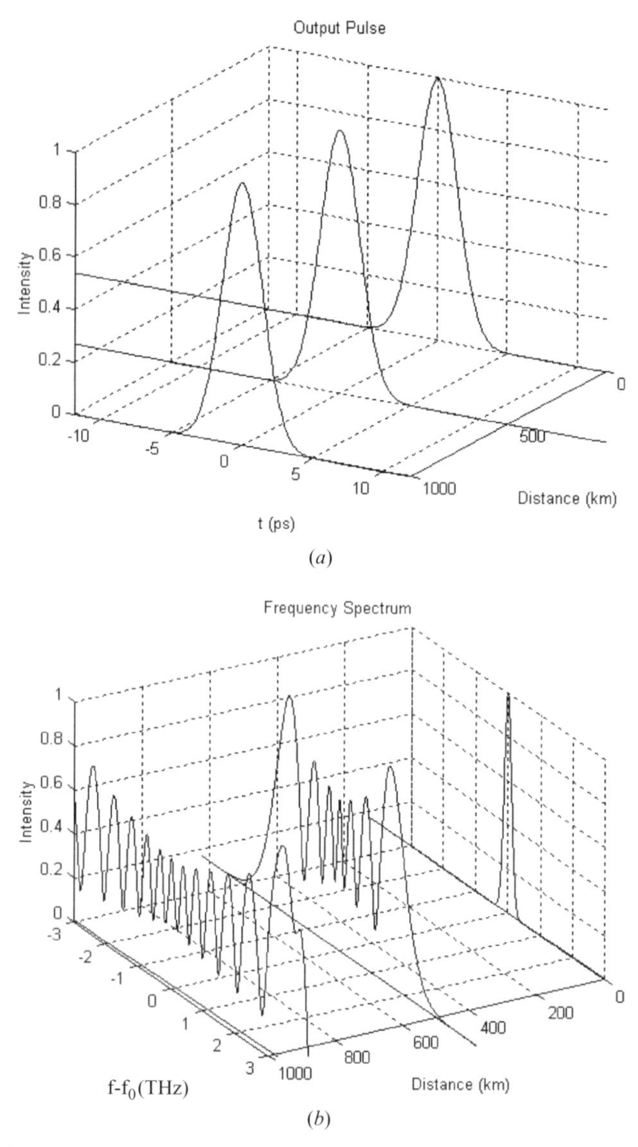

FIGURE 13.7 (*a*) In the presence of nonlinearity alone, the pulse does not broaden; however, its frequency spectrum gets broadened as shown in (*b*).

13.3 SOLITONS

We mentioned in Section 13.2 that the effective dispersion of an optical fiber operating in the positive-dispersion regime gets reduced in the presence of nonlinearity. Under certain conditions of peak power and pulse shape, it is indeed possible to

FIGURE 13.8 Cross-section of a photonic crystal fiber consisting of a series of regularly arranged holes surrounding a very small core. Light confinement in such fibers is very strong, leading to very high intensity levels.

compensate completely for dispersive effects by using nonlinear effects. When this happens, the pulse of light would then propagate without broadening. Such pulses, called *solitons*, are very important for fiber optic communication systems since by eliminating dispersion it is possible to achieve increased bit rates and hence increase the information capacity of the system.

The peak power P_0 of a soliton and the pulse width τ_f are related through the equation

$$P_0 \approx 1.55 \frac{\lambda_0^2 \, |D|}{\pi c \gamma \tau_f^2} \tag{13.8}$$

where τ_f is the full width at half maximum (FWHM) of the soliton pulse, D represents the dispersion coefficient of the fiber, and λ_0 is the wavelength of the soliton. As an

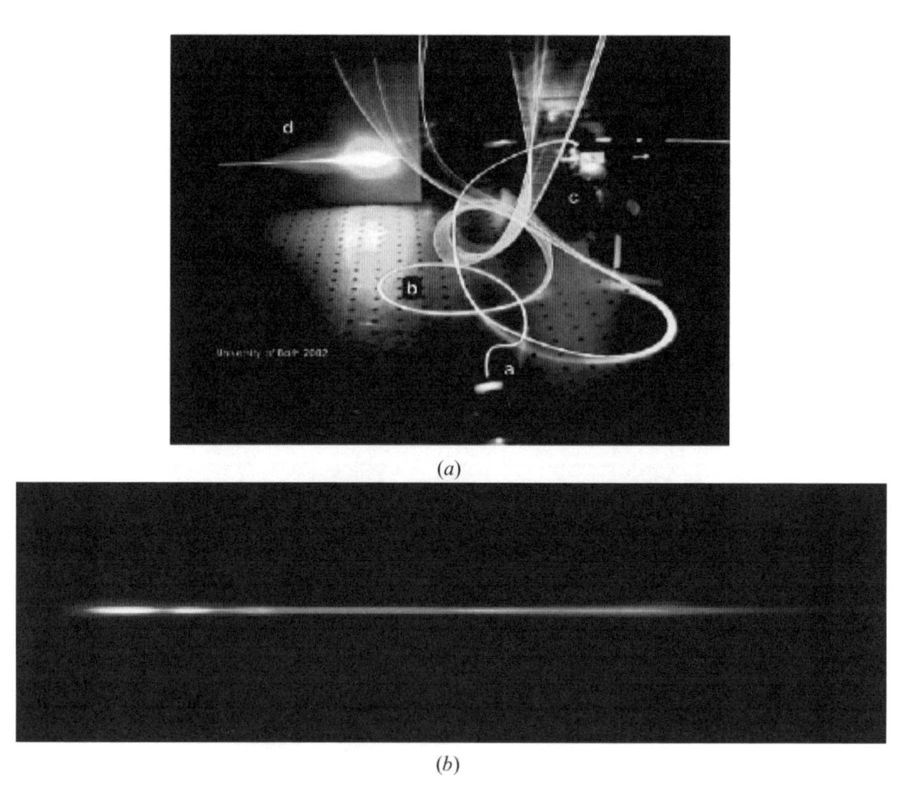

(*a*)

(*b*)

FIGURE 13.9 Supercontinuum generation from a highly nonlinear optical fiber. (Adapted from Russel, 2003.)

example, we have $\tau_f = 10$ ps, $\gamma = 2.4\ \text{W}^{-1}\ \text{km}^{-1}$, $\lambda_0 = 1.55\ \mu\text{m}$, $D = 2$ ps/km·nm, and the required peak power will be $P_0 = 33$ mW. A heuristic derivation of the power required to form a soliton by cancellation of chirping due to dispersion and nonlinearity can be found in Ghatak and Thyagarajan (1998). Figure 13.10 shows input and output soliton pulses as they emerge after propagation through 2000 km of optical fiber. It can be seen that even after propagating through 2000 km, the pulse shows hardly any dispersion.

In the discussion above, we assumed that the fiber is lossless. Since actual fibers have losses, the power in an optical pulse would keep decreasing as it propagates along the fiber. Since nonlinear effects depend on the power in an optical pulse, this would imply that as the pulse propagates, the nonlinear effects would keep decreasing. Hence, if at the input there is enough power for the soliton, as it propagates the power would be insufficient for the formation of a soliton and the pulse would then start to disperse. This problem is partially compensated for by using optical amplifiers periodically along the optical fiber link. The amplification could be carried out using

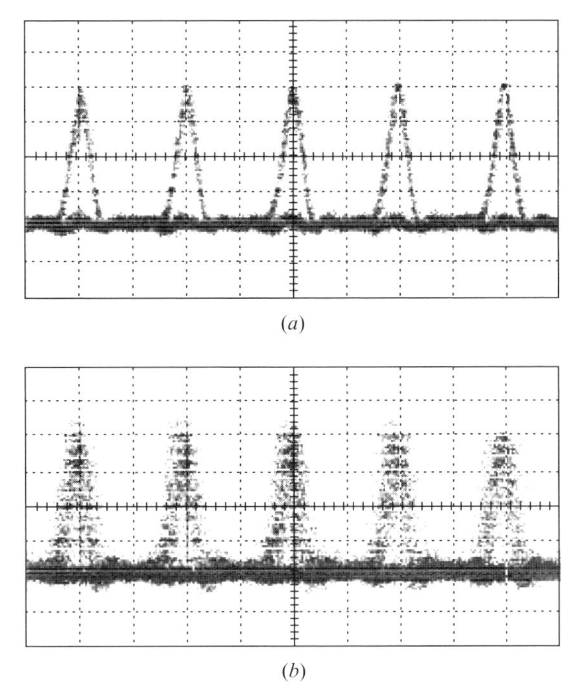

FIGURE 13.10 Soliton propagation over 2000 km of single-mode fiber: (*a*) 0 km; (*b*) 2000 km. (After Toda et al., 1998.)

erbium-doped fiber amplifiers or fiber Raman amplifiers. Soliton pulses are being studied extensively for application to long-distance optical fiber communication.

13.4 CROSS-PHASE MODULATION

In WDM systems, within the fiber there are pulses propagating simultaneously at different wavelengths. In the presence of nonlinear effects, each wavelength would result in a change in refractive index of the fiber, depending on the power carried by that wavelength. If we now consider light beams at two different frequencies propagating simultaneously through the fiber, the change in refractive index brought about by each of the beams will affect the propagation of the other beam. This effect, termed *cross-phase modulation* (XPM), results in *crosstalk*: the output from one channel now depending on the presence or absence of the other channel. Since the signal pulses are random, sometimes there would be overlap between the two signal pulses and sometimes there would be no overlap. When they overlap there would be effects of cross-phase modulation and when there is no overlap, there would be no cross-phase modulation. This random nature results in a random noise of the channels, resulting in a penalty in terms of increased bit error rates.

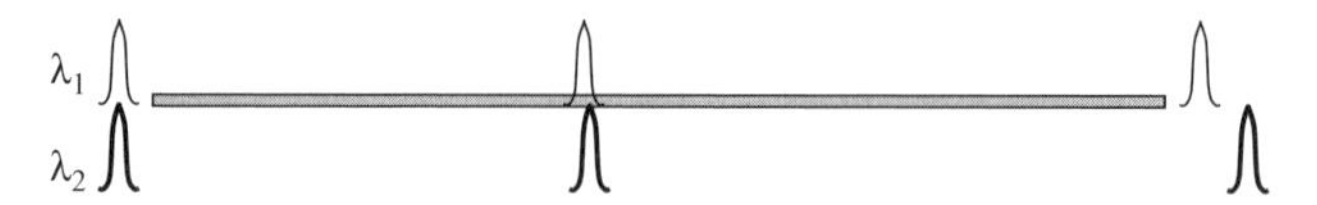

FIGURE 13.11 Cross-phase modulation takes place when pulses at different wavelengths overlap.

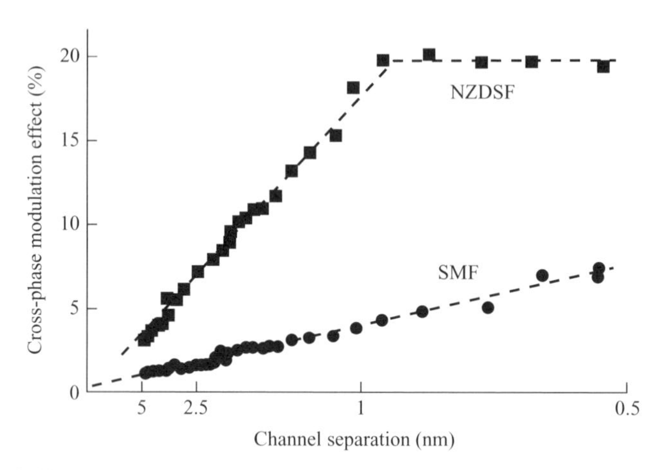

FIGURE 13.12 Variation of cross-phase modulation effect with channel spacing for standard SMF and nonzero dispersion-shifted fiber. (Adapted from Shtaif et al., 2000. Copyright © 2000 IEEE.)

If the signals at both frequencies are pulses, then due to the difference in velocities of the pulses, there would be a walk-off between the two pulses (i.e., if they start together, they will separate as they propagate through the fiber; Fig. 13.11). Nonlinear interaction will take place as long as they overlap physically in the medium. Since smaller dispersion values correspond to smaller difference in velocities (assuming closely spaced wavelengths), the pulses will overlap with each other over longer distance along the fiber. This would lead to stronger XPM effects. Thus, XPM effects would be reduced in fibers possessing larger dispersion, thus favoring the use of fibers with nonzero dispersion values (Fig. 13.12).

13.5 FOUR-WAVE MIXING

Four-wave mixing (FWM) is a nonlinear interaction that occurs in the presence of multiple wavelengths in a fiber, leading to the generation of new frequencies. Let us consider a WDM system with a certain number of input channels operating at equally spaced frequencies. If we denote three of the input channel frequencies as $f_0, f_0 - u$, and $f_0 + u$, the same nonlinearity that led to an intensity-dependent refractive index leads to the generation of many new frequencies. The closely-lying frequencies generated are $f_0 - 3u$, $f_0 - 2u$, $f_0 + 2u$, and $f_0 + 3u$ (Fig. 13.13). This process is referred to as four-wave mixing, due to the interaction among four

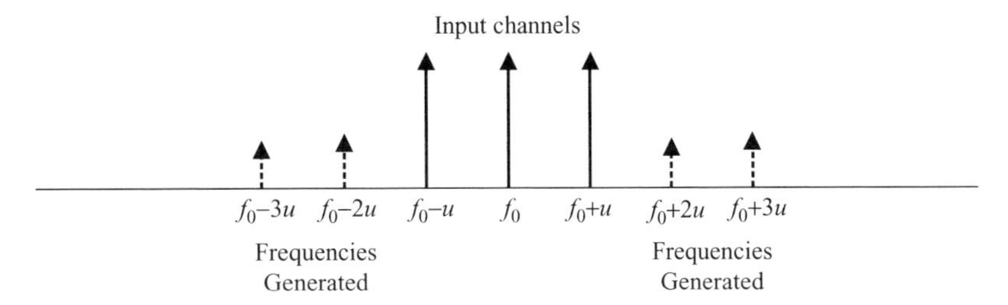

FIGURE 13.13 When three input signals at the frequencies $f_0, f_0 + u$, and $f_0 - u$ propagate through an optical fiber, four-wave mixing generates many new frequencies. Shown are some nearby new frequencies generated by FWM.

different frequencies. In a WDM system carrying multiple channels, FWM can cause the generation of spurious signals at frequencies corresponding to the other channels. Thus, if there are signals present at any frequency corresponding to any of the newly generated frequency, there would be increased noise at these frequencies. This would result in severe crosstalk among the many propagating channels, and thus it is necessary that FWM effects are minimized in WDM systems.

An efficient interaction and hence maximum FWM between the various frequencies takes place when the frequencies lie close to the zero-dispersion wavelength of the fiber. If we assume closely-lying wavelengths near the zero-dispersion wavelength of the fiber and that all wavelengths carry the same power P_{in}, the ratio of the power generated due to FWM to that exiting from the fiber is approximately given by

$$R = \frac{P_g}{P_{out}} = 4\gamma^2 P_{in}^2 L_{eff}^2 \tag{13.9}$$

Thus, using $\gamma = 1.73 \times 10^{-3}\ \text{W}^{-1}\ \text{m}^{-1}$, $L_{eff} = 20$ km, we obtain

$$R = 4.8 \times 10^{-3} P_{in}^2(\text{mW}) \tag{13.10}$$

Thus, if in each channel we have an input power of 1 mW, the FWM-generated output will be about 0.5% of the power exiting in the channel. This gives us the level of crosstalk among the channels created due to FWM.

This is the main problem in using wavelength-division multiplexing (WDM) in dispersion-shifted fibers which are characterized by zero dispersion at the operating wavelength of 1550 nm, as FWM will then lead to crosstalk among various channels. To reduce FWM effects, the dispersion must be nonzero at the wavelength of operation. Thus, dispersion-shifted fibers operating close to the zero-dispersion wavelength will exhibit a significant four-wave mixing problem. This has led to the development of nonzero dispersion-shifted fiber (NZDSF), which has a finite nonzero dispersion of about ±2 ps/km·nm at the operating wavelength. If the frequency spacing between the channels decreases, the dispersion value would need to be increased further for

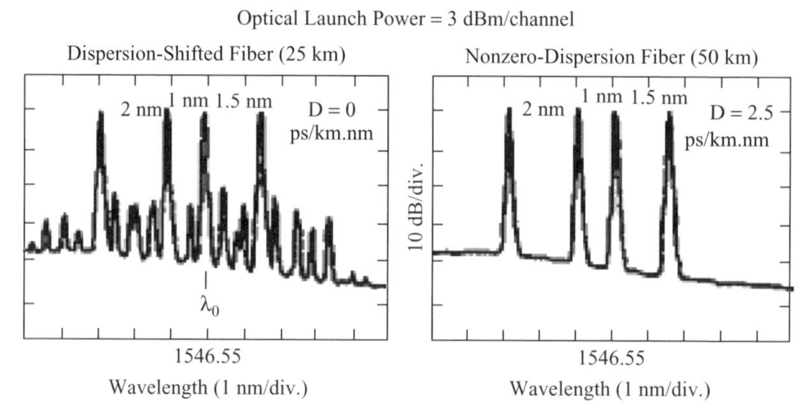

FIGURE 13.14 Four-wave mixing takes place between three input pulses and generates new frequencies if the wavelengths are close to the zero-dispersion wavelength. On the other hand, if the dispersion is finite, FWM effects are greatly reduced. (Adapted from Forghieri et al., 1997.)

reducing the effects of four-wave mixing. Indeed, the standard single-mode fiber (G.652 fiber), which has zero dispersion in the 1310-nm band, has a significant amount of dispersion in the 1550-nm band and thus poses no problems with regard to FWM, even for very small channel spacings.

Figure 13.14 shows the spectrum measured at the output of a 25-km-long dispersion-shifted fiber when four signals, each carrying 3 mW of power at four different wavelengths, are launched simultaneously. Notice the generation of many new frequencies by FWM. The figure also shows that it is possible to reduce the four-wave mixing efficiency by choosing a nonzero value of dispersion.

Since dispersion leads to increased bit error rates in fiber optic communication systems, it is important to have low dispersion. On the other hand, lower dispersion leads to crosstalk due to FWM. This problem can be resolved by noting that FWM depends on the local dispersion value in the fiber, while pulse spreading at the end of a link depends on the overall dispersion in the fiber link. If one chooses a link made up of positive and negative dispersion coefficients, then by an appropriate choice of the lengths of the positive and negative dispersion fibers, it would be possible to achieve a zero total link dispersion while maintaining a large local dispersion. In fiber optic systems this is referred to as *dispersion management*.

Although FWM leads to crosstalk among different wavelength channels in an optical fiber communication system, it can be used for various optical processing functions, such as wavelength conversion, high-speed time-division multiplexing, and pulse compression. For such applications, there is a concerted worldwide effort to develop highly nonlinear fibers with much smaller mode areas and higher nonlinear coefficients. Some of the very novel fibers that have been developed recently include holey fibers, photonic bandgap fibers, and photonic crystal fibers, which are very interesting since they possess extremely small mode effective areas ($\sim$2.5 μm^2 at

1550 nm) and can be designed to have zero dispersion even in the visible region of the spectrum. This is expected to revolutionize nonlinear fiber optics by providing new geometries to achieve highly efficient nonlinear optical processing at lower powers.

13.6 SUPERCONTINUUM GENERATION

Supercontinuum (SC) generation is the phenomenon in which a nearly continuous spectrally broadened output (bandwidth >1000 nm) is produced through nonlinear effects on high-peak-power picosecond and femtosecond pulses. Such broadened spectra find applications in spectroscopy, optical coherence tomography, and as WDM sources for optical communication by slicing the spectrum. Supercontinuum generation in an optical fiber is a very convenient technique since the intensity levels can be maintained high over long interaction lengths by choosing small mode areas, and the dispersion profile of the fiber can be designed appropriately by varying the transverse refractive index profile of the fiber. The spectral broadening that takes place in the fiber is attributed to a combination of various third-order effects, such as SPM, XPM, FWM, and Raman scattering. Since dispersion plays a significant role in the temporal evolution of the pulse, different dispersion profiles have been used in the literature to achieve broadband SC. Some studies have used dispersion-decreasing fibers and dispersion-flattened fibers, while others have used a constant anomalous dispersion fiber followed by a normal dispersion fiber.

Figure 13.15 shows the spectrum of the input and output light pulses from a photonic crystal fiber. The enormous spectral broadening is apparent. Figure 13.16 shows the SC spectrum generated by passing a 25-GHz optical pulse train at 1544 nm generated by a mode-locked laser diode and amplified by an erbium-doped fiber. The

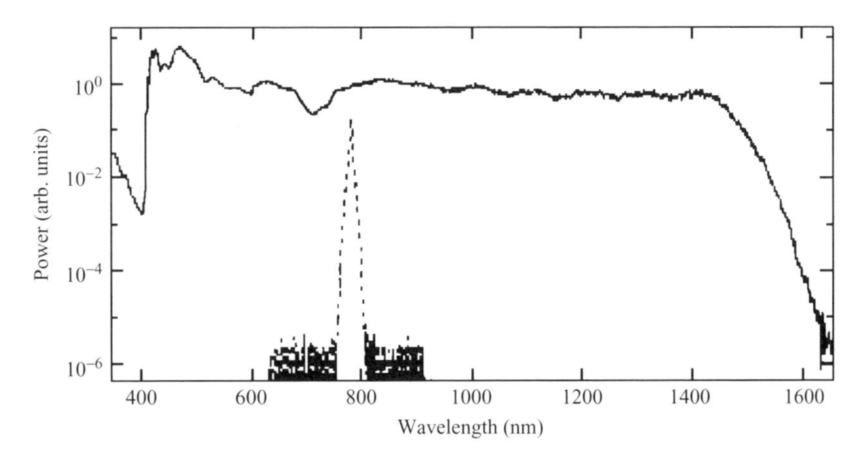

FIGURE 13.15 Spectra of input and output pulses from a photonic crystal fiber. The enormous increase in bandwidth due to nonlinear effects is apparent. (Adapted from Ranka et al., 2000.)

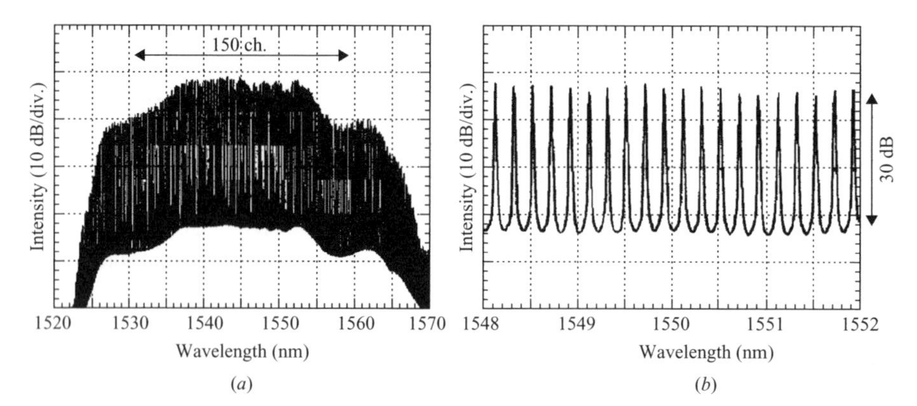

FIGURE 13.16 Supercontinuum spectrum generated by passing a 25-GHz optical pulse train at 1544 nm generated by a mode-locked-laser diode and amplified by an erbium-doped fiber. (After Yamada et al., 2001.)

fiber used for SC generation is a polarization-maintaining dispersion-flattened fiber. The output is a broad spectrum containing more than 150 spectral components at a spacing of 25 GHz with a flattop spectrum over 18 nm. The output optical powers range from -9 to $+3$ dBm. Such sources are being investigated as attractive solutions for dense WDM (DWDM) optical fiber communication systems.

13.7 CONCLUSIONS

The small cross-sectional areas and long interaction lengths of optical fibers leads to significant nonlinear effects. These nonlinear effects ultimately limit the information-carrying capacity of the fiber. At the same time, some nonlinear effects can be used profitably for processing of optical signals.

CHAPTER 14

Optical Fiber Sensors

14.1 INTRODUCTION

Although the most important application of optical fibers is in the field of communication, optical fibers are finding more and more applications in the area of sensing. The use of optical fibers for such applications offers the same advantages as in the field of communication: lower cost, smaller size, rugged, higher accuracy, greater flexibility with multifunctional capabilities, wide range of sensor gauge lengths, and greater reliability. Compared to conventional electrical sensors, such fiber optic sensors are immune to external electromagnetic interference and can also be used in hazardous and explosive environments. A very important attribute of fiber optic sensors is the possibility of having distributed (i.e., measuring over a continuous region) or quasi-distributed (i.e., measuring at a large number of discrete points in some region) sensing geometries, which would otherwise be too expensive or complicated using conventional sensors. Using fiber optic sensors, it is possible to measure almost any external parameter, such as pressure, temperature, electric current, magnetic field, rotation, acceleration, strain, and chemical and biological parameters, with greater precision and speed. These advantages lead to increased integration of such fiber optic sensors into such civil structures as bridges and tunnels, process industries, medical instruments, aircrafts, missiles, and even cars.

Figure 14.1 is a schematic of a fiber optic sensor system. Light from a suitable source is coupled into an optical fiber. The external disturbance modifies some property of the light beam, which is then guided through the optical fiber to a detector. Fiber optic sensors can be classified broadly into two categories: extrinsic and intrinsic. In *extrinsic sensors*, the optical fiber simply acts as a device to transmit and collect light from a sensing element that is external to the fiber. The sensing element responds to the external perturbation, and the change in the characteristics of the sensing element is transmitted by the return fiber for analysis. The optical fiber here plays no role other than transmitting the light beam to and from the sensing region. Such fiber optic sensors are easy to design and fabricate and relatively inexpensive. Examples of such sensors are Doppler anemometers, noncontact vibration measurement, and pressure sensors which find wide applications in automobiles and aerospace, for example.

Fiber Optic Essentials, By K. Thyagarajan and Ajoy Ghatak

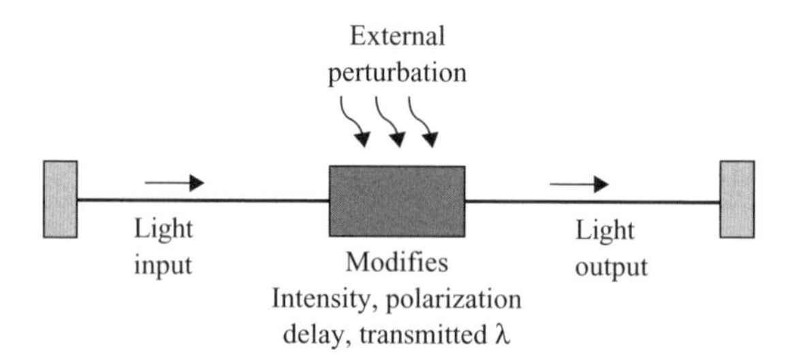

FIGURE 14.1 Schematic of a fiber optic sensing system.

On the other hand, in *intrinsic sensors*, the light beam never leaves the fiber and the physical parameter to be sensed directly alters the properties of the optical fiber, which in turn leads to changes in a characteristic such as intensity, polarization, and phase of the light beam propagating in the fiber leading to sensing. Compared to extrinsic sensors, intrinsic sensors are much more complex, relatively more expensive, but much more sensitive than extrinsic sensors. The sensing signal could be in the form of changes in any of the following properties of the light beam: intensity, phase, polarization, and spectrum, among others. Fiber optic gyroscopes and fiber optic Mach–Zehnder interferometric sensors for acoustic detection are examples of such sensors. The sensor based on fiber Bragg grating discussed in Chapter 11 is an intrinsic fiber optic sensor.

A very large variety of fiber optic sensors have been demonstrated in the laboratory, and some are already installed in real systems. We discuss next some important examples of fiber optic sensors.

14.2 EXTRINSIC FIBER OPTIC SENSORS

Figure 14.2 shows a very interesting fiber optic liquid-level sensor. Light propagating down an optical fiber is totally internally reflected from a small glass prism and couples back to the return fiber. As long as the external medium is air, the angle of

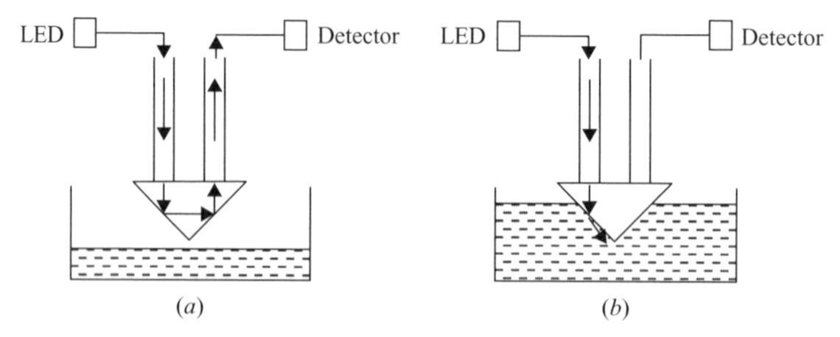

FIGURE 14.2 Fiber optic liquid-level sensor based on changes in the critical angle due to the liquid level moving up: (*a*) total internal reflection; (*b*) no total internal reflection.

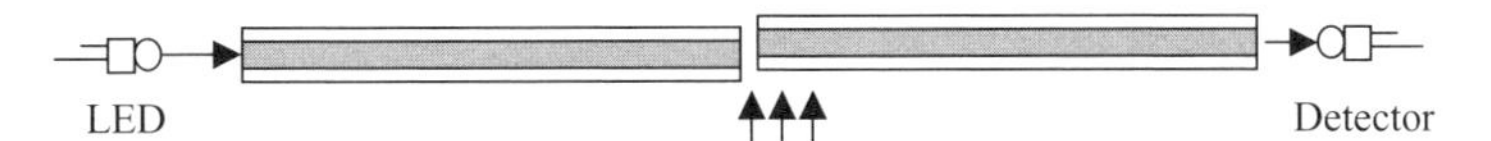

FIGURE 14.3 Change in the transverse alignment between two fibers changes the coupling and hence the power falling on the detector. Such a sensor can detect movements.

incidence inside the prism is greater than the critical angle, and hence light suffers total internal reflection. As soon as the prism comes in contact with a liquid, the critical angle at the prism–liquid interface decreases and the light gets transmitted into the liquid, resulting in a loss of signal. By a proper choice of prism material, such a sensor can be used for sensing levels of various liquids, such as water, gasoline, acids, and oils. One of the greatest advantages of such a sensor is its application to level monitoring of highly inflammable liquids such as petroleum products. Since the sensor does not carry an electrical signal, there is no danger of electrical sparks. Also, being completely dielectric in nature, such a sensor can be used in electrically noisy environments such as close to high-voltage transformers.

Figure 14.3 shows a very simple sensor based on the fact that transmission through a fiber joint depends on alignment of the fiber cores. Light coupled into a multimode optical fiber couples across a joint into another fiber, which is detected by a photodetector. Any deviation of the fiber pair from perfect alignment is sensed immediately by the detector. A misalignment of magnitude equal to the core diameter of the fiber (50 μm) results in zero transmission. A region of about 20% of transverse displacement gives approximately linear output (i.e., the signal is proportional to the displacement). Thus, for a 50-μm core diameter fiber, approximately 10 μm misalignment will be linear. The sensitivity will, of course, increase with a decrease in core diameter, but at the same time, the range of displacements will decrease.

The misalignment between fibers could be caused by various physical parameters, such as acoustic waves and pressure. Thus, if one of the probe fibers has a short free length and the other has is longer, acoustic waves impinging on the sensor would set the fibers into vibration, which would result in a modulation of the transmitted light intensity leading to an acoustic sensor. Using such an arrangement, deep-sea noise levels in the frequency range 100 Hz to 1 kHz and transverse displacements of a few tenths of nanometers have been measured. Using the same principle, any physical parameter leading to relative displacement of the fiber cores can be sensed using this geometry.

Figure 14.4 shows a modification of the sensor in the form of a probe. Here light from an LED coupled into a multimode fiber passes to the probe through a fiber optic

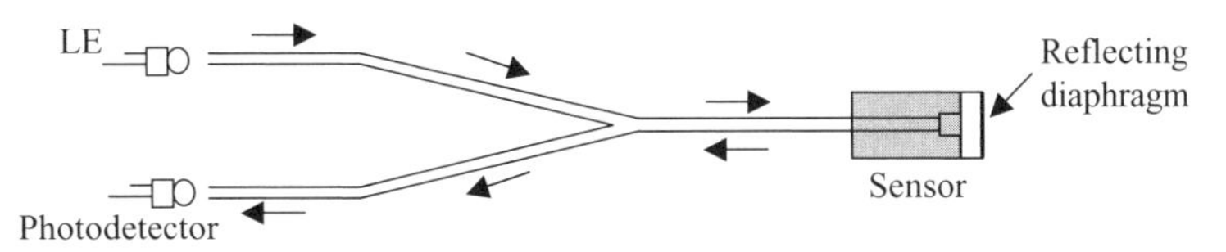

FIGURE 14.4 Light returning to the detector changes as the shape of the reflecting diaphragm changes due to changes in external pressure.

splitter. The probe is in the form of a reflecting diaphragm in front of the fiber, as shown in the figure. Light emanating from the fiber gets reflected by the diaphragm, passes through the splitter again, and is detected by a photodetector. Any changes in the external pressure causes the diaphragm to bend, leading to a change in the power coupled into the fiber. Such sensors can be built to measure pressure variations in medical as well as other applications requiring monitoring operating pressures of up to 4 M Pa ($\simeq$ 600 pounds per square inch). Such a device can indeed be used, for example, in the measurement of pressure in the arteries, bladder, and urethra. Several experiments have shown a very good correlation between this sensor and other conventional sensors.

If the diaphragm at the output is removed and the light beam is allowed to fall on the sample, light that is reflected or scattered is again picked up by the fiber and detected and processed by the detector. By analyzing the returning optical beam, information as to the physical and chemical properties of the sample such as blood can be obtained. Thus, if the scattering takes place from flowing blood, the scattered light beam is shifted in frequency due to the Doppler effect. (The *Doppler effect* is the apparent shift in the frequency of a wave as observed by an observer when the source or the observer has relative motion. You must have noticed the falling frequency of the whistle of a train as it approaches and recedes from you.) The faster the blood cells are moving, the larger will be the change in frequency. By measuring the shift in frequency of the light beam, the blood flow rate can be estimated. By a spectroscopic analysis of the returning optical signal, one can estimate the oxygen content in the blood. One of the most important advantages of using optical fibers in this process is that they do not provoke any adverse response from the immune system, they are more durable, flexible, and potentially safer than alternatives.

14.3 INTRINSIC FIBER OPTIC SENSORS

In intrinsic sensors, the light beam never leaves the fiber and the physical parameter changes some characteristic of the propagating light beam, which is sensed at the output. Among the many intrinsic sensors, here we discuss four examples: bend and microbend sensors, fiber optic current sensors, Mach–Zehnder interterometric fiber sensors, and the fiber optic gyroscope. In Chapter 11 we discussed sensors based on fiber Bragg gratings, which are also intrinsic sensors.

Bend and Microbend Sensors

When an optical fiber is bent, a portion of the propagating light beam along the bend is incident at angles less than the critical angle and thus suffer from attenuation. Thus, bending a fiber results in loss, and the smaller the bending radius, the greater is the loss of light. Bend-induced loss can be used for sensing purposes. Figure 14.5 shows a pressure sensor that is used to measure load distribution. It consists of a pair of arrays of fibers placed at right angles to each other as shown in the figure. Light is coupled into all the fibers in the array, and when a load is applied, the loss of each fiber is measured. From the loss distribution it is possible to estimate the bend

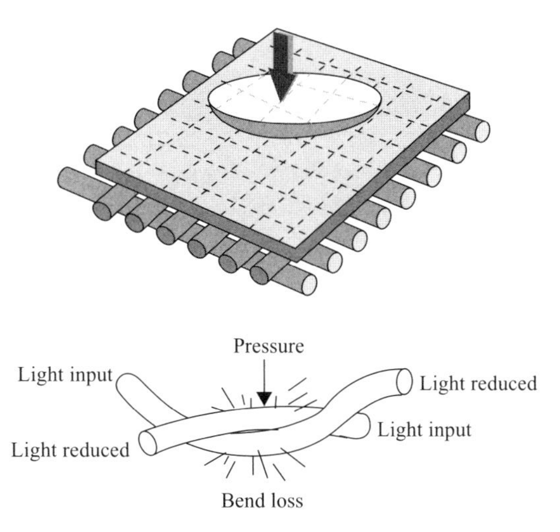

FIGURE 14.5 Fiber optic sensor based on a cross array of fibers. Any load will lead to bending of the fibers, with a resulting loss of power. Measuring the loss of power in each fiber leads to a knowledge of the distribution of strain across the area. (Adapted from Wang et al., 2005.)

distribution and thus the load distribution. Figure 14.6 shows the bend distribution measured when a load of 6.6 N is applied to the fiber sensor.

In Chapter 5 we discussed microbending, which is a series of random and small bends along the fiber. This leads to coupling between various modes in a multimode fiber and between the core and the cladding modes in a single-mode fiber. Both lead

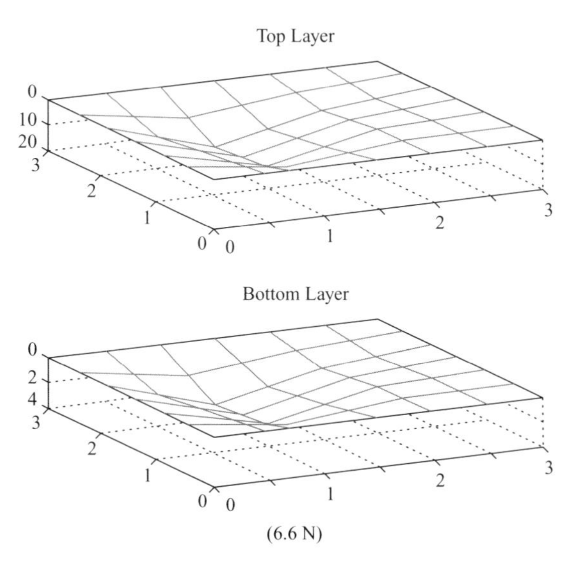

FIGURE 14.6 Force distribution due to load of 6.6 N applied at one point as measured by the fiber optic sensor shown in Fig. 14.5. (Adapted from Wang et al., 2005.)

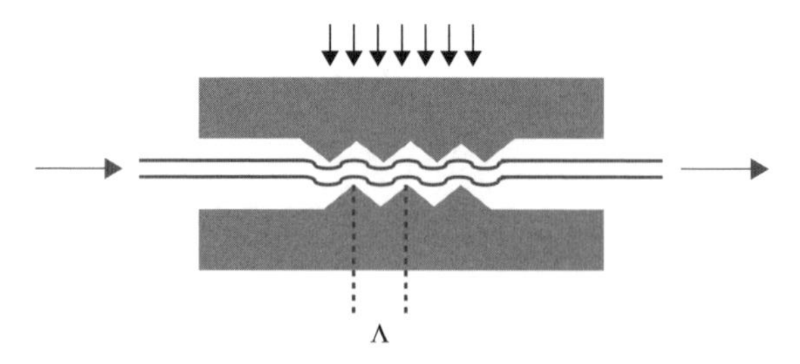

FIGURE 14.7 Microbend fiber optic sensor in which microbends induce transmission losses.

to increased losses in the fiber due to microbending. This effect can be used for sensing by intentionally introducing periodic microbends in the fiber and choosing an appropriate spatial period of the microbend to lead to maximum loss (Fig. 14.7). The mechanism of coupling between various modes is similar to the coupling brought about by fiber gratings. In the case of fiber gratings, it is the periodic refractive index modulation along the core that induces coupling between chosen modes, while in the case of microbending it is the periodic undulation of the axis of the fiber that leads to coupling. The spatial period responsible for coupling various modes is the same as for fiber gratings.

In the case of multimode parabolic index fibers, the propagation constants of various guided modes are equally spaced, with a spacing given by

$$\Delta\beta = \frac{\sqrt{2\Delta}}{a} \tag{14.1}$$

where $\Delta\beta$ is the difference in propagation constants between adjacent modes, a is the core radius, and

$$\Delta = \frac{n_1^2 - n_2^2}{2n_1^2} \approx \frac{n_1 - n_2}{n_1} \tag{14.2}$$

with n_1 being the refractive index of the core along the axis and n_2 the refractive index of the cladding. For coupling by periodic perturbations in the fiber, the spatial frequency of the periodic perturbation should be equal to the difference in propagation constants of the modes that need to be coupled. Hence for maximum coupling, the spatial period Λ of the microbend should be

$$\frac{2\pi}{\Lambda} = \Delta\beta$$

or

$$\Lambda = \frac{2\pi a}{\sqrt{2\Delta}} \tag{14.3}$$

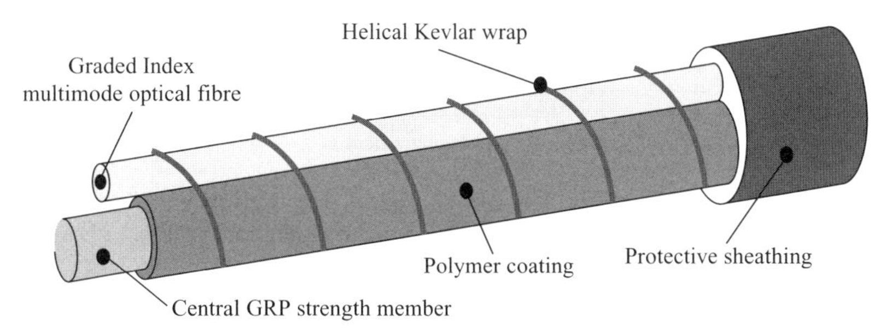

FIGURE 14.8 Hydrocarbon sensor using microbends induced by swelling of polymer due to absorption of hydrocarbon. (Adapted from MacLean et al., 2003.)

Assuming typical values of $\Delta = 0.2$ and a core radius of 50 μm, we obtain 0.7 mm for the period required for maximum coupling. The loss induced due to this microbending depends on the amplitude of the microbends (just like the amplitude of the refractive index modulation in a fiber grating), and monitoring the loss induced by the coupling, it is possible to sense changes in external parameters causing the microbends.

A very interesting sensor based on this idea is the microbend hydrocarbon sensor. It consists of a parabolic index optical fiber placed against a strength member coated with a polymer. A helical Kevlar wrap is wound around at the right pitch (Fig. 14.8). When the polymer gets exposed to hydrocarbon it results in swelling of the polymer, which in turn induces microbending on the fiber. This microbend induces loss in the fiber at that location. Using an optical time-domain reflectometer (OTDR) it is possible to estimate the location of the microbend and hence of any leakage of hydrocarbon. Figure 14.9 shows a typical OTDR trace of such a sensor where 0.5-m sections of the fiber were exposed to petrol.

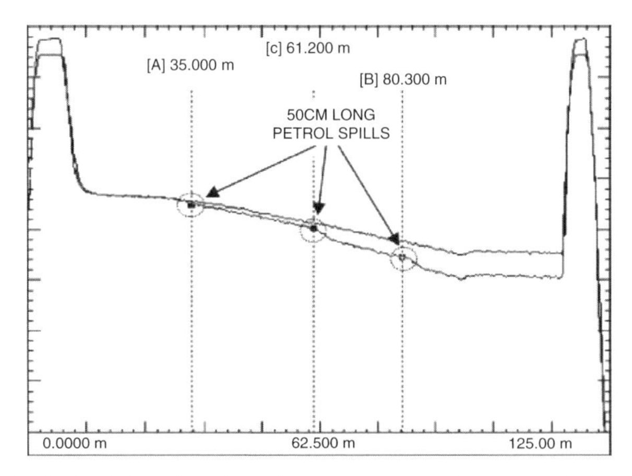

FIGURE 14.9 Optical time-domain reflectometry trace of a sensor. It shows two points of loss, due to microbends induced by absorption of petrol. (Adapted from MacLean et al., 2003.)

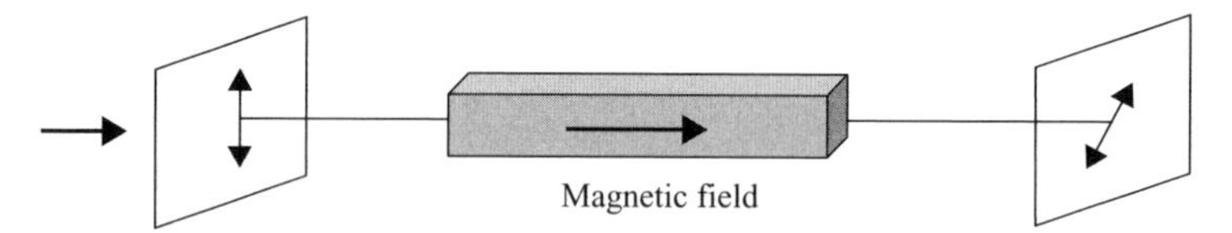

FIGURE 14.10 Faraday effect. A linearly polarized light with its electric field oriented vertically changes its orientation as it propagates through a medium in which a magnetic field is applied along the direction of propagation.

Fiber Optic Current Sensors

The fiber optic current sensor is based on the Faraday effect. When a linearly polarized light beam propagates through media such as ordinary glass, the state of polarization remains the same along the propagation. Now, if a magnetic field is applied along or opposite to the direction of propagation of the light beam in the medium, it is found that the polarization direction of the propagating light beam undergoes a rotation (Fig. 14.10; see also Fig. 12.10). This effect of rotation of the plane of polarization of a linearly polarized light beam is referred to as the *Faraday effect*. The angle of rotation is proportional to the length traversed and to the applied magnetic field. It also depends on the material through a constant called the *Verdet constant*.

Figure 14.11 shows a typical experimental setup for the measurement of current in the conductor using the Faraday effect. It consists of a single-mode fiber (about 400 m long) wound helically around the current-carrying conductor. The magnetic field associated with the current (which encircles the current-carrying wire) results in the rotation of the state of polarization of the light propagating through the fiber.

Referring to Fig. 14.11, if the current passing through the conductor is I and if there are N turns of fiber around the current-carrying conductor, the angle of rotation of the plane of polarization is given by

$$\theta = VNI \tag{14.4}$$

where V is the Verdet constant, which for silica has a value 2.64×10^{-4} degree/A. By measuring the rotation of the plane of polarization of the light beam as it propagates through an optical fiber, it is possible to measure the magnetic field. The rotation in

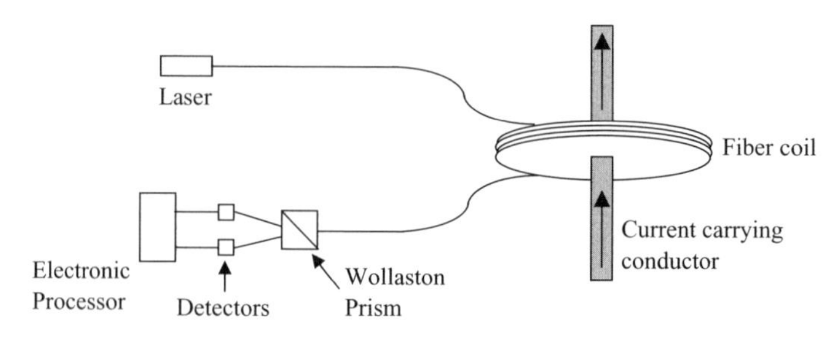

FIGURE 14.11 Fiber optic current sensor based on the Faraday effect.

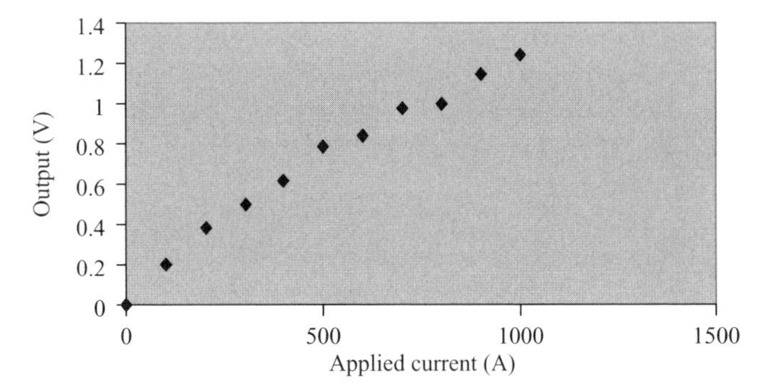

FIGURE 14.12 Measurements from a fiber optic current sensor.

the plane of polarization is detected by passing the output through a suitable device. Since a current-carrying conductor produces a magnetic field, using this effect it is indeed possible to measure current. As an example, if we consider 30 turns of fiber wound around a current-carrying conductor, for a current of 10 A the rotation is 7.9×10^{-2} degree. It is possible to measure such a rotation by proper signal processing.

The rotation of the plane of polarization is indeed independent of the shape of the loop as well as any current sources lying outside the loop and is dependent only on the current through the conductor. The final signal after processing by the electronic circuits is proportional to the current in the wire. Figure 14.12 shows a typical output signal from the sensor as a function of current passing through the conductor. Currents greater than tens of kiloamperes are measurable using this principle.

One major issue in such a sensor is the effect of bending the fiber in loops. When a fiber is bent, it becomes birefringent (i.e., light waves polarized in the plane of the bend and perpendicular to the plane of the bend do not propagate with the same velocity). The smaller the loop radius, the larger is this difference. In the presence of this effect the polarization state of the propagating light is further modified, and this results in loss of sensitivity of the sensor. One method to reduce the effect of bending is to introduce twisting in the fiber. Using twist rates of a few rotations per meter, the effects of bend-induced birefringence can be reduced.

Mach-Zehnder Interferometric Sensors

One of the most sensitive arrangements for a fiber optic sensor is the Mach–Zehnder (MZ) interferometric sensor arrangement shown in Fig. 14.13. Light from a laser is passed through a fiber optic coupler, which splits the incoming light beam into two equal-amplitude beams in the two single-mode fiber arms. One of the arms, called the *reference arm*, is kept isolated from any external perturbation. The fiber in this arm is sometimes coated with a material to make it insensitive to the parameter of measurement. The other arm, called the *sensing arm*, is exposed to the perturbation that is to be measured. After traversing through the two arms, the light beams are made to recombine at the output coupler. The output light is detected by a photodetector

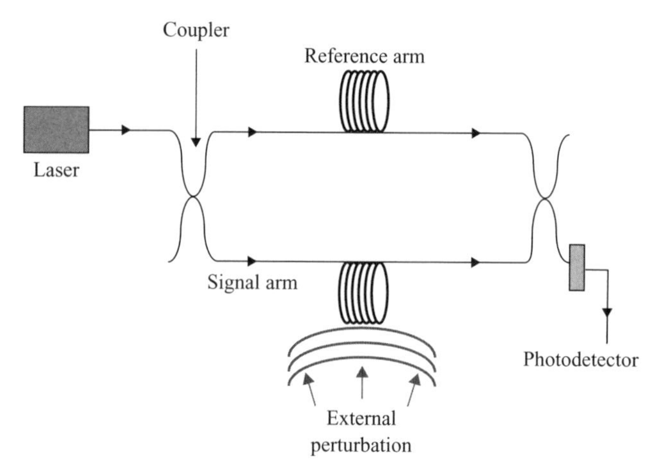

FIGURE 14.13 Fiber optic Mach–Zehnder interferometric sensor. Changes in phase between light beams arriving at the output coupler due to external perturbation on the sensing arm cause changes in intensity at the output.

and processed electronically. In this sensor, the two fiber arms behave as the two paths of an interferometer, and hence the output light power falling on the detector would depend on the phase difference between the beams as they enter the output coupler. If the two fibers are of exactly equal lengths, the entire input light beam appears in the lower fiber and no light comes out of the upper fiber. Any external parameter, such as temperature or pressure, affects the sensing fiber by changing either the refractive index or the length of the arm, thus changing the phase difference between the two beams as they enter the output coupler. This results in a change in the output power falling on the detector (Fig. 14.14):

$$P_d = P_0 \cos^2 \frac{\Delta\phi}{2} \tag{14.5}$$

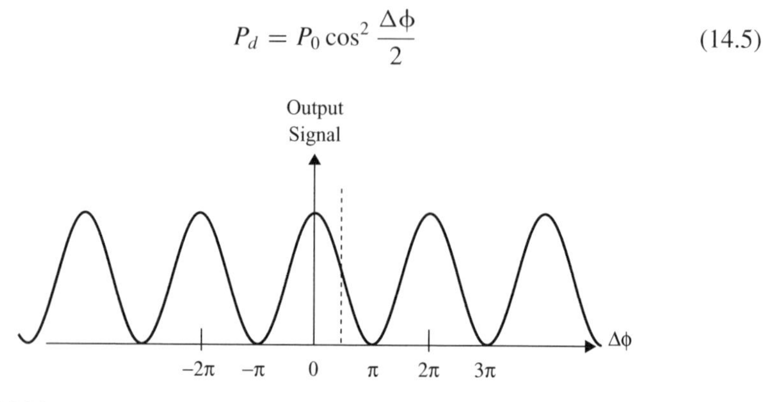

FIGURE 14.14 Intensity variation at the output of the interferometer as a function of the phase difference between the two arms. The vertical dashed line shows the quadrature operating point for maximum sensitivity.

where P_0 is the input power, P_d is the power falling on the detector, and $\Delta\phi$ is the phase change brought about by the external perturbation. Processing the output leads to measurement of the parameter.

Since the power exiting is maximum around the point given by $\Delta\phi = 0$, operating the sensor around this point would lead to very poor sensitivity, since a given change of $\Delta\phi$ would lead to a very small change in the output power. The most sensitive point of operation of the sensor is shown by the vertical dashed line in Fig. 14.14, also referred to as the *quadrature point*, since it corresponds to $\Delta\phi = \pi/2$. The sensor can be biased externally to operate at this point. If *d* represents the phase change induced by the external measurand, and if the sensor is operated at the quadrature point, we can write

$$\Delta\phi = \frac{\pi}{2} + \delta \tag{14.6}$$

Assuming that $\delta \ll \pi$, substituting from Eq. (14.6) in Eq. (14.5), we obtain

$$P_d = \frac{P_0}{2}(1 - \delta) \tag{14.7}$$

Thus in this case the power variations are linearly related to the phase change δ.

With sensitive detection schemes it is possible to measure phase changes of 10^{-6} radian. Any external perturbation causing such a small phase change can be measured by the sensor. The MZ sensor is extremely sensitive to external perturbations. For example, when someone whispers, a pressure wave is created in the air. When such a pressure wave hits the fiber, it leads to very minute changes in the refractive index of the fiber. Using a sensing arm of about 100 m, one can even detect such minute changes and hence listen to someone whispering. Such Mach–Zehnder interferometric sensors are now used as hydrophones for underwater sound detection in marine applications and to detect presence of submarines. One important advantage of such an sensor is the possibility of configuring them as omnidirectional (i.e., sensor capable of picking sound emanating from any direction) or highly directional (sensitive to sound waves from only a very specific direction) sensors.

MZ sensors can be used to sense any physical parameter, such as temperature, strain, or magnetic field, which can cause changes in the phase of the propagating light beam. The pressure and temperature sensitivities of the MZ interferometric sensor are

$$\frac{\Delta\phi}{L\,\Delta P} \approx -5 \times 10^{-5}\ \mathrm{rad/Pa \cdot m} \tag{14.8}$$

$$\frac{\Delta\phi}{L\,\Delta T} \approx 100\ \mathrm{rad/K \cdot m} \tag{14.9}$$

Example 14.1 Let us consider a MZ interferometer with a 100-m-long sensing arm consisting of a coated optical fiber with a sensitivity of 3×10^{-4} rad/Pa·m. Sound at the threshold of hearing falling on the sensing arm would induce a phase change of

$$\Delta\phi = (3 \times 10^{-4})(2 \times 10^{-5})(100) = 6 \times 10^{-7} \text{ rad}$$

Fiber Optic Rotation Sensor: The Fiber Optic Gyroscope

One of the most important fiber optic sensors is the fiber optic gyroscope, which is capable of measuring rotation rates of platforms. The FOG is a relatively low cost device without moving parts that demonstrates a better lifetime than that of conventional gyroscopes. Thus FOGs are rapidly replacing conventional mechanical gyros for many applications.

The principle of operation of the fiber optic gyroscope is based on the Sagnac effect. Figure 14.15 shows a Sagnac interferometer in which a laser beam is split into two equal parts at a beamsplitter, with one portion propagating clockwise and the other anticlockwise through a loop made up of three mirrors, as shown. After propagating through the loop, the two beams recombine at the same beamsplitter and emerge from the output arm. Since the clockwise and anticlockwise beams travel through the same path, they would normally take the same time to go around the loop and hence would arrive with the same phase at the beamsplitter. Now, if this loop rotates about an axis perpendicular to the loop, we can imagine the following situation: In the time the clockwise and anticlockwise beams traverse the loop, the beamsplitter would have moved slightly, as shown in Fig. 14.16. This would imply that the clockwise beam would have to travel a slightly longer distance to reach the beamsplitter, and the anticlockwise beam, a slightly shorter distance. This would result in a phase difference between the two beams, leading to a change in intensity at the output from the beamsplitter. Thus, by measuring the intensity at the output,

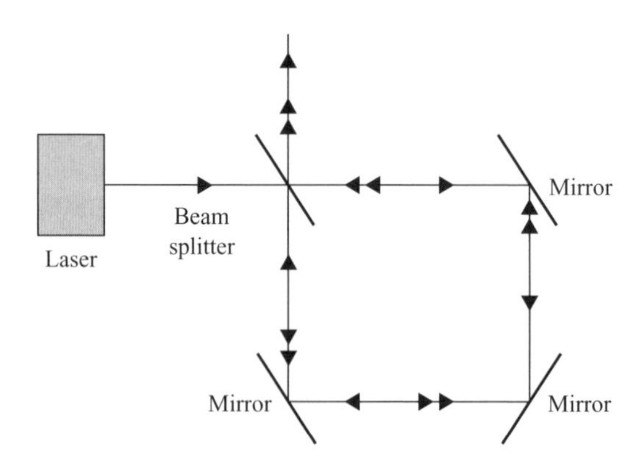

FIGURE 14.15 Sagnac interferometer.

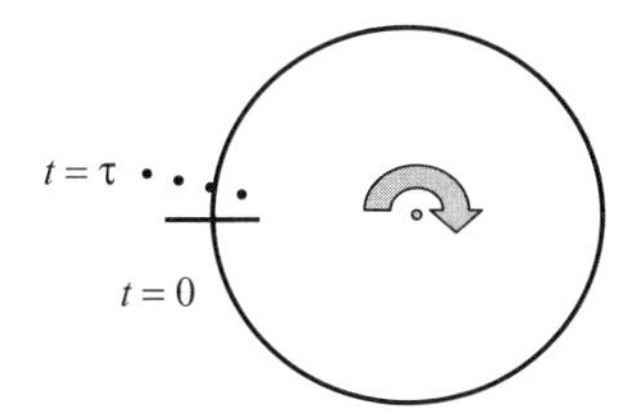

FIGURE 14.16 If the loop rotates in the clockwise direction, by the time the clockwise and counterclockwise beams return, the beamsplitter would have moved, and thus the two beams would have slightly different durations.

it is possible to measure the rotation rate and thus act like a gyroscope. The phase difference between the clockwise and anticlockwise beams is given by

$$\Delta\phi = \frac{8\pi N A\Omega}{c\lambda_0} \tag{14.10}$$

where N is the number of fiber turns in the loop, A the area of one fiber loop, Ω the angular velocity of rotation, and λ_0 the free-space wavelength of light.

Figure 14.17 shows a typical fiber optic gyroscope in which the light from the source (usually, a superluminiscent diode with a spectral width of about 30 nm) is split into clockwise and anticlockwise beams using a fiber optic coupler. After propagating through the loop, the beams recombine, using the same coupler. The returning beams are then coupled out into a detector using another fiber optic coupler. The loop is configured out of polarization maintaining single-mode optical fiber so that the beams maintain their polarization states as they propagate through the loop. The setup shown ensures that the clockwise and counterclockwise beams travel through identical paths and that any effect other than rotation would affect the two beams in a similar fashion and would not lead to any spurious signal. This can be seen from the fact that at the output coming out of the coupler connected to the loop, the two interfering beams would not have propagated through identical paths since the clockwise beam would correspond to straight paths through the coupler, whereas the

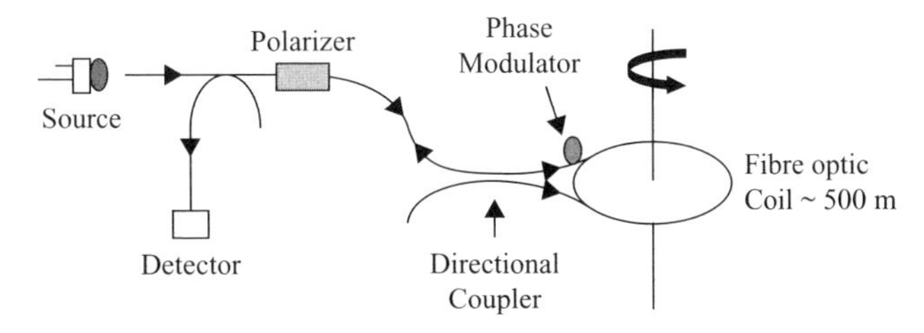

FIGURE 14.17 Fiber optic gyroscope based on the Sagnac effect.

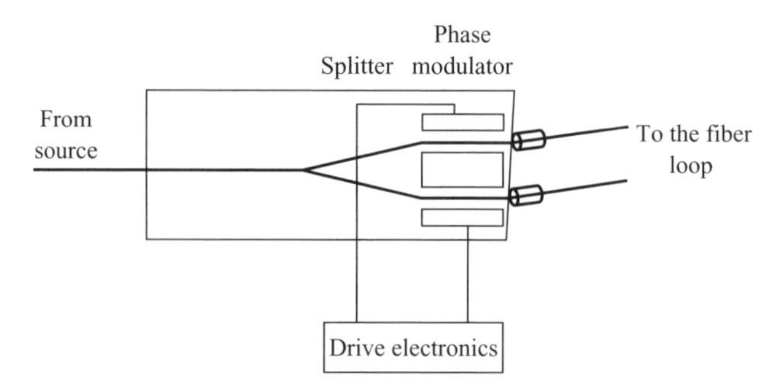

FIGURE 14.18 Integrated optic chip that integrates the functions of coupler, polarizer, and phase modulator on a single chip in lithium niobate.

anticlockwise beam would correspond to light that has crossed over twice through the coupler. This problem is eliminated using the signal exiting the second coupler placed close to the source. The phase modulator provided within the fiber loop is used to introduce an artificial nonreciprocal phase shift between the clockwise and anticlockwise beams so as to enable efficient processing of signal.

As seen from Eq. (14.10) the phase difference between the two beams is directly proportional to the rotation rate, the number of fiber turns, and the area of the loop. As a typical example, if we consider a coil of diameter 10 cm having 1500 turns, then due to the Earth's rotation (15° per hour), the phase difference introduced between the clockwise and counter clockwise beams is a tiny 0.0001 radian. This corresponds to a flight-time difference between the two beams of about 4×10^{-20} s! The FOG is capable of detecting rotation rates up to 0.001° per hour and is thus a very sensitive device for the measurement of rotation rates.

In an actual fiber optic gyroscope the total fiber length that can be used is limited by the loss in the fiber, and the radius of the loop is constrained by other factors, such as bend loss and birefringence due to bend. Some of the components used in a fiber optic gyroscope can be integrated into a single optical chip, leading to what is referred to as a *FOG chip*. Figure 14.18 shows a schematic of an integrated optic chip on lithium niobate substrate in which the coupler, phase modulator, and polarizer are integrated into a single chip. Such chips can lead to efficient, low-cost solutions of FOGs.

One of the great advantages of the Sagnac interferometer is that the sensor gives no signal for reciprocal stimuli (i.e., stimuli that act in an identical fashion on both beams). Thus, a change of temperature affects both beams (clockwise and anticlockwise) equally and so produces no change in the output.

Fibre optic gyros capable of measuring from 0.001° to 100° per hour are being made. Applications include navigation of aircraft, spacecraft, missiles, manned and unmanned platforms, antenna piloting and tracking, and a compass or north finder. Various applications require FOGs with different sensitivities: automobiles require about 10 to 100° per hour, attitude reference for airplanes require 1° per hour, and precision inertial navigation requires gyros with 0.01 to 0.001° per hour. Boeing 777 aircraft use an inertial navigation system that has both ring laser gyroscopes and FOGs.

An interesting application is for automobile navigation. The autogyro provides information about the direction and distance traveled and the vehicle's location, which is shown on a monitor in the car. Thus, the rider can navigate himself through a city. Luxury cars from Toyota and Nissan sold in Japan have FOGs as a part of their onboard navigation system. The biggest manufacturer of such gyros is Hitachi of Japan, which manufactures about 3000 gyros per month.

REFERENCES AND SUGGESTED READING

Agarwal, G. P. (1989). *Nonlinear Fiber Optics*, Academic Press, San Diego, CA.

Auguste, J. L., R. Jindal, J. M. Blondy, M. Clapeau, J. Marcou, B. Dussardier, G. Monnom, D. B. Ostrowsky, B. P. Pal, and K. Thyagarajan (2000). −1800 ps/km·nm chromatic dispersion at 1.55 μm in a dual concentric core fiber, *Electronics Letters*, **36**, 1689–1691.

Becker, P. C., N. A. Olsson, and J. R. Simpson (1999). *Erbium Doped Fiber Amplifiers*, Academic Press, San Diego, CA.

Bergano, N. S. (2005). Wavelength division multiplexing in long haul transoceanic transmission systems, *Journal of Lighwave Technology*, **23**, 4125.

Bromage, J. (2004). Raman amplification for fiber communication systems, *Journal of Lightwave Technology*, **22**, 79–93.

Desurvire, E. (1994). *Erbium Doped Fiber Amplifiers*, Academic Press, San Diego, CA.

Emori, Y., and S. Namiki (1999). 100 nm bandwidth flat gain Raman amplifiers pumped and gain-equalized by 12-wavelength-channel WDM high power laser diodes, *Proceedings of the Optical Fiber Communication Conference*, Post-deadline Paper PD19.

Forghieri, F., R. W. Tkach, and A. R. Chraplyvy (1997). Fiber nonlinearities and their impact on transmission systems, in *Optical Fiber Telecommunications*, Vol. **IIIA**, I. P. Kaminow and T. L. Koch (Editors), Academic Press, San Diego, CA.

Gambling, W. A. (1986). Glass, light, and the information revolution, Ninth W.E.S. Turner Memorial Lecture, *Glass Technology*, **27**(6), 179.

Gates, W. (1996). *The Road Ahead*, Penguin Books, New York.

Gebremicheal, Y. M., W. Li, B. T. Meggitt, W. J. O. Boyle, K. T. V. Grattan, B. McKinley, L. F. Broswell, K. A. Ames, S. E. Aasen, B. Tynes, Y. Fonjallaz, and T. Triantafillou (2005). A field deployable, multiplexed Bragg grating sensor system used in an extensive highway bridge monitoring evaluation test, *IEEE Sensors Journal*, **5**, 510–519.

Ghatak, A., and K. Thyagarajan (1998). *Introduction to Fiber Optics*, Cambridge University Press, Cambridge.

Goyal, I. C., R. K.Varshney, and A. K. Ghatak (2003). Design of a small residual dispersion fiber and a corresponding dispersion compensating fiber for DWDM systems, *Optical Engineering*, **42**, 977–980.

Guy, M., and Y. Painchaud (2004). Fiber Bragg gratings: a versatile approach to dispersion compensation, *Photonics Spectra*, August.

Hecht, J. (1999). *City of Light*, Oxford University Press, New York.

Hecht, J. (2004). A new nanotwist for unclad optical fibers, *Optics and Photonics News*, April, p. 21.

Fiber Optic Essentials, By K. Thyagarajan and Ajoy Ghatak

Islam, M. N. (2002). Raman amplifiers for telecommunications, *IEEE Selected Topics in Quantum Electronics*, **8**, 548.

Jacob, J. M., E. A. Golovchenko, A. N. Pilipetskii, G. M. Carter, and C. R. Menyuk (1997). 10 Gb/s transmission of NRZ over 10000 km and solitons over 13500 km error free in the same dispersion managed system, *IEEE Photonics Technology Letters*, **9**, 1412.

Jeong, Y., J. K. Sahu, D. N. Payne, and J. Nilsson (2004). Ytterbium-doped large core fiber laser with 1.36 kW continuous wave output power, *Optics Express*, **12**, 6088.

Johnson, J. E.. L. J. P. Ketelsen, D. A. Ackerman, L. Zhang, M. S. Hybertson, K. G. Glogovsky, C. W. Lentz, W. A. Asous, C. L. Reynolds, J. M. Geary, K. K. Kamath, C. W. Ebert, M. Park, G. J. Przybylek, R. E. Leibenguth, S. L. Broutin, J. W. Stayt, K. F. Dreyer, L. J. Peticolas, R. L. Hartman, and T. L. Koch (2001). Fully stabilized electroabsorption-modulated tunable DBR laser transmitter for long haul optical communications, *IEEE Journal of Selected Topics in Quantum Electronics*, **7**, 168.

Kakkar, C., and K. Thyagarajan (2005). High gain Raman amplifier with inherent gain flattening and dispersion compensation, *Optics Communications*, **250**, 77–83.

Kao, C. K., and G. A. Hockham (1966). Dielectric fiber surface waveguides for optical frequencies. *IEE Proceedings*, **133**, 1151.

Kashyap, R. (1999). *Fiber Bragg Gratings*, Academic Press, San Diego, CA.

Kritler, D. (2003). Laser optimized multimode fiber presents standardized testing challenges, *Lightwave*, September.

Lucas-Lectin, G., T. Avignon, and L. Jacubowiez (2005). An optical time domain reflectometry set up for laboratory work at École Supérieure d'Optique, ETOP05, Marseille.

MacLean, A., C. Moran, W. Johnstone, B. Culshaw, D. Marsh, P. Parker (2003). Detection of hydrocarbon fuel spills using a distributed fibre optic sensor, *Sensors and Actuators* A, **109**, 60–67.

Maclean, D. J. H. (1996). *Optical Line Systems*, Wiley, Chichester, West Sussex, England.

Matthijsse, P., G. Kyut, F. Gooijer, F. Achten, F. Freund, L. Molle, C. Caspar, T. Rosin, A. Beling, and T. Eckhardt (2006). Multimode fiber enabling 40 Gbit/s multi-mode transmission over distances >400 m, *Proceedings of the Optical Fiber Communications Conference*, Paper OWI-13.

McGhan, D., C. Laperle, A. Savehenko, Chuandong Li, G. Mak, M. O'Sullivan (2005). 5120 km RZ-DPSK transmission over G652 fiber at 10 Gb/s with no optical dispersion compensation, *Technical Digest, Optical Fiber Communication Conference*, Vol. 5; OFC/NFOEC, Vol. 6.

Miya, T., Y. Terunama, T. Hosaka, and T. Miyashita (1979). An ultimate low loss single mode fiber at 1.55 μm, *Electronics Letters*, **15**, 106.

Mizuno, K., Y. Nishi, Y. Mimura, Y. Iida, H. Matsuura, D. Yoon, O. Aso, T. Yamamoto, T. Toratani, Y. Ono, and A. Yo (2000). Development of etalon-type gain-flattening filter, *Furukawa Review*, No. 19, 53–58.

Moriyama, T., O. Fukuda, K. Sanada, K. Inada, T. Edahvio, and K. Chida (1980). Ultimately low OH content V.A.D. optical fibers, *Electronics Letters*, **16**, 689.

Nagel, S. (1989). Optical fiber: the expanding medium, *IEEE Circuits Devices*, **36**, March.

Pal, B. P. (2006). *Guided Wave Optical Components and Devices*, Elsevier, Amsterdam, The Netherlands.

Palai, P., M. N. Satyanarayan, M. Das, K. Thyagarajan, and B. P. Pal (2001). Characterization and simulation of long period gratings using electric discharge, *Optics Communications*, **193**, 181–185.

Pondillo, P. (2001). Multimode fiber for use with laser sources, White Paper WP4119, Corning Glass Works, Corning, NY.

Pone, E., X. Daxhelet, and S. Lacroix (2004). Refractive index profile of fused-fiber couplers cross section, *Optics Express*, **12**, 1036–1044.

Ramachandran, S. (2005). Dispersion tailored few-mode fibers: A versatile platform for in-fiber photonic devices, *J. Lightwave Technology*, **23**, 3426–3443.

Ramaswami, R., and K. N. Sivarajan (1998). *Optical Networks: A Practical Perspective*, Morgan Kaufmann, San Francisco, CA.

Ranka, J. K., R. S. Windeler, and A. J. Stentz (2000). Visible continuum generation in air silica microstructure optical fibers with anomalous dispersion at 800 nm, *Optics Letters*, **25**, 25–27.

Rottwitt, K. (2005). Lecture notes on "*Raman Amplification Using Optical Fibers*," CGCRI, Kolkata, India.

Russel, P. (2003). Photonic crystal fibers, *Science*, **299**, January 17, p. 358.

Sakabe, I., H. Ishikawa, H. Tanji, Y. Terasawa, M. Itou, and T. Ueda (2005). *SEI Technical Review*, **59**, January, p. 32.

Shtaif, M., M. Eiselt, and L. D. Garret (2000). Cross phase modulation distortion measurements in multispan WDM systems, *IEEE Photonics Technology Letters*, **12**, 88–90.

Sundar, V. C., A. D. Yablon, J. L. Grazul, M. Ilan, and J. Aizenberg (2003). Fiber optical features of a glass sponge, *Nature*, **424**, 899.

Thyagarajan, K., and C. Kakkar (2004). Novel fiber design for flat gain Raman amplification using single pump and dispersion compensation in S-band, *Journal of Lightwave Technology*, **22**, 2279–2286.

Toda, H., Y. Inada, Y. Kodama, and A. Hasegawa (1998). 10 Gbit/s optical soliton transmission experiment in a comb like dispersion profiled fiber loop, *Proceedings of the 24th European Conference on Optical Communication* (ECOC'98), MoC09, p. 101.

Upadhyaya, B. N., A. Kuruvilla, S. Kher, M. R. Shenoy, K. Thyagarajan, and T. P. S. Nathan (2005). Development and analysis of a highly-efficient 10W single-mode Yb-doped CW fiber laser, *Proceedings of ICOL*, Dehradun, India.

Venkataraman, G. (1994). *Journey into Light: Life and Science of C. V. Raman*, Penguin Books, New York.

Wang W., W. R. Ledoux, B. J. Sangeorzan, and P. G. Reinhall (2005). A shear and plantar pressure sensor based on fiber optic bend loss, *Journal of Rehabilitation Research and Development*, **42**, 315.

Yamada, E., H. Takara, T. Ohara, K. Sato, T. Morioka, K. Jinguji, M. Itoh, and M. Ishi (2001). 150 channel supercontinuum CW optical source with high SNR and precise 25 GHz spacing for 10 Gbit/s DWDM systems, *Electronics Letters*, **37**, 304–306.

Zeigler, K. E., R. J. Lascola, and L. L. Tovo (2003). Fiber optic laser Raman spectroscopy sensor, Westinghouse Savannah River Company, South Carolina (http://sti.srs.gov/fulltext/tr2003284/2003284.pdf).

INDEX

Fiber Optic Essentials, By K. Thyagarajan and Ajoy Ghatak